Humphry Davy

The Elementary Nature of Chlorine

Papers

Humphry Davy

The Elementary Nature of Chlorine
Papers

ISBN/EAN: 9783337026998

Printed in Europe, USA, Canada, Australia, Japan

Cover: Foto ©berggeist007 / pixelio.de

More available books at **www.hansebooks.com**

Alembic Club Reprints—No. 9.

THE

ELEMENTARY NATURE

OF

CHLORINE.

PAPERS BY

HUMPHRY DAVY

(1809-1818).

Edinburgh:

PUBLISHED BY THE ALEMBIC CLUB.

Edinburgh Agent:
WILLIAM F. CLAY, 18 TEVIOT PLACE.

London Agents:
SIMPKIN, MARSHALL, HAMILTON, KENT & CO., LTD.

1902.

PREFACE.

THIS reprint contains the important contributions of Humphry Davy on the properties and nature of "muriatic acid" and "oxymuriatic acid," which led to the re-adoption of Scheele's view of the relation in which these bodies stand to each other as against that advanced by Berthollet. It has been considered advisable to include the early Papers on muriatic acid, so as to shew more clearly the change which took place in Davy's views, and the causes which led to it. That oxymuriatic acid *might be* a simple body seems to have been first suggested by Gay Lussac and Thenard, in 1809, but they themselves rejected this view at the time, and it was left for Davy to show its correctness. The discussion which these Papers aroused, known as the "Chlorine Controversy," was contributed to by Humphry Davy only to a slight extent, his brother, Dr. John Davy, taking the chief part in defending the theory.

In collecting Davy's contributions to any one subject it is difficult to give all that is wanted without giving unnecessary matter as well, since many of the larger Papers deal with more than one subject. This is especially the case with the Bakerian Lectures. In the present case nothing has been omitted from any Paper unless it was without direct bearing upon the subject of the reprint. All excisions are indicated by asterisks.

H. M.

A List of the various Papers will be found on page 78.

*The Bakerian Lecture. An Account of some new analytical Researches on the Nature of certain Bodies, particularly the Alkalies, Phosphorus, Sulphur, Carbonaceous Matter, and the Acids hitherto undecompounded; with some general Observations on Chemical Theory.**

Read December 15, 1808.

1. Introduction.

IN the following pages, I shall do myself the honour of laying before the Royal Society, an account of the results of the different experiments, made with the hopes of extending our knowledge of the principles of bodies by the new powers and methods arising from the applications of electricity to chemistry, some of which have been long in progress, and others of which have been instituted since their last session.

The objects which have principally occupied my attention, are the elementary matter of ammonia, the nature of phosphorus, sulphur, charcoal, and the diamond, and the constituents of the boracic, fluoric, and muriatic acids.

Amongst the numerous processes of decomposition, which I have attempted, many have been successful; and from those which have failed, some new phenomena have

* [From " Philosophical Transactions " for 1809, vol. 99, pp. 39–104. Part reprinted, pp. 39, 40, 91–103.]

usually resulted which may possibly serve as guides in future inquiries. On this account, I shall keep back no part of the investigation, and I shall trust to the candour of the Society for an excuse for its imper-fection.

The more approaches are made in chemical inquiries towards the refined analysis of bodies, the greater are the obstacles which present themselves, and the less perfect the results.

All the difficulties which occur in analysing a body, are direct proofs of the energy of attraction of its con-stituent parts. In the play of affinities with respect to secondary compounds even, it rarely occurs that any perfectly pure or unmixed substance is obtained ; and the principle applies still more strongly to primary com-binations.

The first methods of experimenting on new objects likewise are necessarily imperfect ; novel instruments are demanded, the use of which is only gradually acquired, and a number of experiments of the same kind must be made, before one is obtained from which correct data for conclusions can be drawn.

*　　*　　*　　*　　*

8. *Analytical Experiments on Muriatic Acid.*

I have made a greater number of experiments upon this substance, than upon any of the other subjects of research that have been mentioned ; it will be impossible to give any more than a general view of them within the limits of the Bakerian lecture.

Researches carried on some years ago, and which are detailed in the Journals of the Royal Institution, shewed that there were little hopes of decomposing muriatic acid, in its common form, by VOLTAIC electricity. When aqueous solution of muriatic acid is acted upon, the water alone is decomposed; and the VOLTAIC electrization of the gas affords no indications of its decomposition; and merely seems to shew, that this elastic fluid contains much more water than has been usually suspected.

I have already laid before the Society, an account of some experiments made on the action of potassium on muriatic acid. I have since carried on the same processes on a larger scale, but with precisely similar results.

When potassium is introduced into muriatic acid gas, procured from muriate of ammonia and concentrated sulphuric acid, and freed from as much moisture as muriate of lime is capable of attracting from it, it immediately becomes covered with a white crust, it heats spontaneously, and by the assistance of a lamp, acquires in some parts the temperature of ignition, but does not inflame. When the potassium and the gas are in proper proportions, they both entirely disappear; a white salt is formed, and a quantity of pure hydrogene gas evolved, which equals about one third of the original volume of the gas.

By eight grains of potassium employed in this way, I effected the absorption of nearly twenty two cubical inches of muriatic acid gas; and the quantity of hydrogene gas produced was equal to more than eight cubical inches.

The correspondence between the quantity of hydrogene generated in cases of this kind, and by the action of potassium upon water, combined with the effects of ignited charcoal upon muriatic acid gas, by which a quantity of inflammable gas is produced equal to more than one third of its volume; seemed to shew, that the phenomena merely depended upon moisture combined with the muriatic acid gas.*

To determine this point with more certainty however, and to ascertain whether or no the appearance of the hydrogene was wholly unconnected with the decomposition of the acid, I made two comparative experiments on the quantity of muriate of silver, furnished by two equal quantities of muriatic acid, one of which had been converted into muriate of potash by the action of potassium, and the other of which had been absorbed by water; every care was taken to avoid sources of error; and it was found that there was no notable difference in the weight of the results.

There was no proof then, that the muriatic acid had been decompounded in these experiments; and there was every reason to consider it as containing in its common aeriform state, at least one third of its weight of water; and this conclusion we shall find warranted by facts, which are immediately to follow.

I now made a number of experiments, with the hopes of obtaining the muriatic acid free from water.

I first heated to whiteness, in a well luted porcelain retort, a mixture of dry sulphate of iron, and muriate of lime which had been previously ignited; but a few cubic inches of gas only, were obtained, though the mixture was in the quantity of several ounces; and this gas contained sulphureous acid. I heated dry muriate of lime, mixed both with phosphoric glass and dry boracic

---

* When the VOLTAIC spark is taken continuously, by means of points of charcoal in muriatic acid gas over mercury, muriate of mercury is rapidly formed, a volume of inflammable gas, equal to one third of the original volume of the muriatic acid gas appears. The acid gas enters into combination with the oxide of mercury, so that water enough is present in the experiment to form oxide sufficient to absorb the whole of the acid.

acid, in tubes of porcelain, and of iron, and employed
the blast of an excellent forge; but by neither of these
methods was any gas obtained, though when a little
moisture was added to the mixtures, muriatic acid was
develloped in such quantities, as almost to produce
explosions.

The fuming muriate of tin, *the liquor of Libavius,* is
known to contain dry muriatic acid. I attempted to
separate the acid from this substance, by distilling it with
sulphur and with phosphorus; but without success. I
obtained only triple compounds, in physical characters,
something like the solutions of phosphorus, and sulphur
in oil, which were non-conductors of electricity, which
did not redden dry litmus paper, and which evolved
muriatic acid gas with great violence, heat, and ebullition
on the contact of water.

I distilled mixtures of corrosive sublimate and sulphur,
and of calomel and sulphur; when these were used in
their common states, muriatic acid gas was evolved; but
when they were dried by a gentle heat, the quantity was
exceedingly diminished, and the little gas that was
generated gave hydrogene by the action of potassium.
During the distillation of corrosive sublimate and sulphur,
a very small quantity of a limpid fluid passed over.
When examined by transmitted light, it appeared yellowish
green. It emitted fumes of muriatic acid, did not redden
dry litmus paper, and deposited sulphur by the action
of water. I am inclined to consider it as a modification of the substance discovered by Dr. THOMSON, in
his experiments on the action of oxymuriatic acid on
sulphur.

M. M. GAY LUSSAC and THENARD* have mentioned,
that they endeavoured to procure dry muriatic acid by
distilling a mixture of calomel and phosphorus, and that

* The Moniteur before quoted [May 27, 1808].

they obtained a fluid which they consider as a compound of muriatic acid, phosphorus, and oxygene. In dis tilling corrosive sublimate with phosphorus, I had a similar result, and I obtained the substance in much larger quantities, than by the distillation of phosphorus with calomel.

As oxymuriatic acid is slightly soluble in water, there was reason to suppose, reciprocally that water must be slightly soluble in this gas ; I endeavoured therefore to procure dry muriatic acid, by absorbing the oxygene from oxymuriatic acid gas by substances, which when oxygenated, produced compounds possessing a strong affinity for water. Phosphorus, it is well known, burns in oxymuriatic acid gas ; though the results of this com- bustion, I believe, have never been minutely examined. With the hopes of procuring muriatic acid gas, free from moisture, I made the experiment. I introduced phos- phorus into a receiver having a stop-cock, which had been exhausted, and admitted oxymuriatic acid gas. As soon as the retort was full, the phosphorus entered into combustion, throwing forth pale white flames. A white sublimate collected in the top of the retort, and a fluid as limpid as water, trickled down the sides of the neck. The gas seemed to be entirely absorbed, for when the stop-cock was opened, a fresh quantity of oxymuriatic acid, nearly as much as would have filled the retort, entered.

The same phenomenon of inflammation again took place, with similar results. Oxymuriatic acid gas was admitted till the whole of the phosphorus was consumed.

Minute experiments proved, that no gaseous muriatic acid had been evolved in this operation, and the muriatic acid was consequently to be looked for either in the white sublimate, or in the fluid which had formed in the neck of the retort.

The sublimate was in large portions, the fluid only in the quantity of a few drops. I collected by different processes, sufficient of both for examination.

The sublimate emitted fumes of muriatic acid when exposed to air. When brought in contact with water, it evolved muriatic acid gas, and left phosphoric acid, and muriatic acid, dissolved in the water. It was a non-conductor of electricity, and did not burn when heated; but sublimed when its temperature was about that of boiling water, leaving not the slightest residuum. I am inclined to regard it as a combination of phosphoric, and muriatic acid in their dry states.

The fluid was of a pale greenish yellow tint, and very limpid; when exposed to air, it rapidly disappeared, emitting dense white fumes which had a strong smell differing a little from that of muriatic acid.

It reddened litmus paper in its common state, but had no effect upon litmus paper which had been well dried, and which was immediately dipped into it. It was a non-conductor of electricity. It heated when mixed with water, and evolved muriatic acid gas. I consider it as a compound of phosphorous acid, and muriatic acid, both free from water.*

Having failed in obtaining uncombined muriatic acid in this way, I performed a similar process with sulphur, but I was unable to cause it to inflame in oxymuriatic acid gas. When it was heated in it, it produced an orange coloured liquid, and yellow fumes passed into the neck of the retort, which condensed into a greenish yellow

---

* I attempted to obtain dry muriatic acid likewise from the phosphuretted muriatic acid of M. M. GAY LUSSAC and THENARD, by distilling it in retorts containing oxygene gas, and oxymuriatic acid gas. In the first case, the retort was shattered by the combustion of the phosphorus, with a violent explosion. In the second, compounds, similar to those described above, were formed.

fluid. By repeatedly passing oxymuriatic acid through this fluid, and distilling it several times in the gas, I rendered it of a bright olive colour, and in this case it seemed to be a compound of dry sulphuric, and muriatic acid, holding in solution a very little sulphur. When it was heated in contact with sulphur, it rapidly dissolved it, and then became of a bright red colour, and when saturated with sulphur, of a pale golden colour.* No permanent aeriform fluid was evolved in any of these operations, and no muriatic gas appeared, unless moisture was introduced.

As there seemed little chance of procuring uncombined muriatic acid, it was desirable to ascertain what would be the effects of potassium upon it in these singular com-pounds.

When potassium was introduced into the fluid, gene-rated by the action of phosphorus on corrosive sublimate, at first it slightly effervesced, from the action of the liquid on the moist crust of potash surrounding it ; but the metal soon appeared perfectly splendid, and swim-ming on the surface. I attempted to fuse it by heating the fluid, but it entered into ebullition at a temperature below that of the fusion of the potassium ; indeed the mere heat of the hand was sufficient for the effect. On examining the potassium, I found that it was combined at the surface with phosphorus, and gave phosphuretted hydrogene by its operation upon water.

I endeavoured, by repeatedly distilling the fluid from potassium in a close vessel, to free it from phosphorus, and in this way I succeeded in depriving it of a consider-able quantity of this substance.

I introduced ten or twelve drops of the liquid, which

---

* All these substances seem to be of the same nature as the singular compound, the sulphuretted muriatic acid, discovered by Dr. THOMSON, noticed in page 9.

had been thus treated, into a small plate glass retort, containing six grains of potassium; the retort was exhausted after having been twice filled with hydrogene, the liquid was made to boil, and the retort kept warm till the whole had disappeared as elastic vapour. The potassium was then heated by the point of a spirit lamp; it had scarcely melted, when it burst into a most brilliant flame, as splendid as that of phosphorus in oxygene gas, and the retort was destroyed by the rapidity of combustion.

In other trials made upon smaller quantities after various failures, I was at last able to obtain the results; there was no proof of the evolution of any permanent elastic fluid during the operation. A solid mass remained of a greenish colour at the surface, but dark gray in the interior. It was extremely inflammable, and often burnt spontaneously when exposed to air; when thrown upon water, it produced a violent explosion, with a smell like that of phosphuretted hydrogene. In the residuum of its combustion there was found muriate of potash, and phosphate of potash.

I endeavoured to perform this experiment in an iron tube, hoping that if the muriatic acid was decomposed in the process, its inflammable element, potassium and phosphorus, might be separated from each other by a high degree of heat; but in the first part of the operation the action was so intense, as to produce a destruction of the apparatus, and the stop-cock was separated from the tube with a loud detonation.

I heated potassium in the vapour of the compound of muriatic and phosphoric acid; but in this case, the inflammation was still more intense, and in all the experiments that I have hitherto tried, the glass vessels have been either fused or broken; the solid residuum has however appeared to be of the same kind as that I have just described.

The results of the operation of the sulphuretted compounds containing muriatic acid free from water upon potassium, are still more extraordinary than those of the phosphuretted compounds.

When a piece of potassium is introduced into the substance that distils over during the action of heated sulphur upon oxymuriatic acid, it at first produces a slight effervescence, and if the volume of the potassium considerably exceeds that of the liquid, it soon explodes with a violent report, and a most intense light.

I have endeavoured to collect the results of this operation, by causing the explosion to take place in large exhausted plate glass retorts; but, except in a case in which I used only about a quarter of a grain, I never succeeded. Generally the retort, though connected with the air pump at the time, was broken into atoms; and the explosion produced by a grain of potassium, and an equal quantity of the fluid, has appeared to me considerably louder than that of a musket.

In the case in which I succeeded in exploding a quarter of a grain, it was not possible for me to ascertain if any gaseous matter was evolved ; but a solid compound was formed of a very deep gray tint, which burnt, throwing off bright scintillations, when gently heated, which inflamed when touched with water, and gave most brilliant sparks, like those thrown off by iron in oxygene gas.

Its properties certainly differed from those of any compound of sulphur and potassium that I have seen : whether it contains the muriatic basis must however be still a matter of inquiry.

There is, however, much reason for supposing, that in the singular phenomena of inflammation and detonation that have been described, the muriatic acid cannot be entirely passive : and it does not seem unfair to infer, that the transfer of its oxygene and the production of a novel substance, are connected with such effects, and that the highly inflammable nature of the new compounds, partly depends upon this circumstance. I am still pursuing the inquiry, and I shall not fail immediately to communicate to the Society, such results as may appear to me worthy of their attention.

9. Some general Observations, with Experiments.

*   *   *   *   *

The experiments that I have detailed on the acids, offer some new views with respect to the nature of acidity. That a compound of muriatic acid with oxide of tin or phosphorus should not redden vegetable blues, might be ascribed to a species of neutralization, by the oxide or inflammable body; but the same reasoning will not apply to the dry compounds which contain acid matter only, and which are precisely similar as to this quality. Let a piece of dry and warm litmus paper be moistened with the compound of muriatic and phosphorous acid, it perfectly retains its colour. Let it then be placed upon a piece of moistened litmus paper, it instantly becomes of a bright red, heats and devellopes muriatic acid gas.

All the fluid acids that contain water are excellent conductors of electricity, in the class called that of imperfect conductors; but the compounds to which I have just alluded, are non-conductors in the same degree as oils, with which they are perfectly miscible. When I first examined muriatic acid, in its combinations free from moisture, I had great hopes of decomposing them by electricity; but there was no action without contact of the wires, and the spark seemed to separate no one of their constituents, but only to render them gaseous. The circumstance likewise applies to the boracic acid, which is a good conductor as long as it contains water; but which, when freed from water and made fluid by heat, is then a non-conductor.

The alkalies and the earthy compounds, and the oxides, as dry as we can obtain them, though non-conductors when solid, are, on the contrary, all conductors when rendered fluid by heat.

When muriatic acid, existing in combination with phosphorous or phosphoric acid, is rendered gaseous by the action of water, the quantity of this fluid that disappears, at least equals from one third to two fifths of the weight of the acid gas produced; a circumstance that agrees with the indications given by the action of potassium.*

I attempted to procure a compound of dry muriatic and carbonic acids, hoping that it might be gaseous, and that the two acids might be decomposable at the same time by potassium. The process that I employed was by passing corrosive sublimate in vapour through charcoal ignited to whiteness; but I obtained a very small quantity of gas, which seemed to be a mixture of common muriatic acid gas and carbonic acid gas; a very minute portion of running mercury only was obtained, by a long continuation of the process; and the slight decomposition that did take place, I am inclined to attribute to the production of water, by the action of the hydrogene of the charcoal upon the oxygene of the oxide of mercury.†

* Page 13.

† These facts and the other facts of the same kind, explain the difficulty of the decomposition of the metallic muriates in common processes of metallurgy. They likewise explain other phenomena in the agencies of muriatic salts. In all cases when a muriatic salt is decomposed by an acid, and muriatic acid gas set free, there appears to be a double affinity, that of the acid for the basis, and of the muriatic acid for water; pure muriatic acid does not seem capable of being displaced by any other acid.

In mixing muriatic acid gas with carbonic acid, or oxygene, or hydrogene, the gases being in their common states, as to moisture, there was always a cloudiness produced; doubtless owing to the attraction of their water to form liquid muriatic acid.

On fluoric acid gas no such effect was occasioned. This fact, at first view, might be supposed to shew, that the hydrogene evolved by the action of potassium upon fluoric acid gas, is owing to water in actual combination with it, like that in muriatic acid gas, and which may be essential to its elastic state; but it is more probable, from the smallness of the quantity, and from the difference of the quantity in different cases, that the moisture is merely in that state of diffusion or solution in which it exists in gases in general, though from the disposition of water to be deposited in this acid gas in the form of an acid solution, it must be either less in quantity, or in a less free state, so as to require for its exhibition much more delicate hygrometrical tests.

*   *   *   *   *

Davy.

New analytical Researches on the Nature of certain Bodies, being an Appendix to the Bakerian Lecture for 1808.* †

---

*   *   *   *   *

IV. Further Inquiries respecting muriatic Acid.

The experiments on muriatic acid, which I have already had the honour of laying before the Society, shew that the ideas which had been formerly entertained respecting the difference between the muriatic acid and the oxymuriatic acid are not correct. They prove that muriatic acid gas is a compound of a substance, which as yet has never been procured in an uncombined state, and from one third to one-fourth of water, and that oxymuriatic acid is composed of the same substance, (free from water) united to oxygene. They likewise prove, that when bodies are oxydated in muriatic acid gas, it is by a decomposition of the water contained in that substance, and when they are oxydated in oxymuriatic acid, it is by combination with the oxygene in that body, and in both cases there is always a union of the peculiar unknown substance, the dry muriatic acid with the oxydated body.

---

* The account of the principal facts respecting the action of potassium on ammonia, in this communication, were read before the Royal Society, February 2, 1809. The paper was ordered to be printed March 16, 1809. At that time, having stated to the Council that I had since made some new experiments on this matter, and on the subjects discussed in the Bakerian Lecture for 1808, I received permission to add them to the detail of the former observations for publication.

† [From " Philosophical Transactions" for 1809, vol. 99, pp 450–470. Part reprinted, pp. 468–470.]

Of all known substances belonging to the class of acids, the dry muriatic acid is that which seems to possess the strongest and most extensive powers of combination. It unites with all acid matters that have been experimented upon, except carbonic acid, and with all oxides (including water), and all inflammable substances that have been tried, except those which appear to be elementary, carbonaceous matter and the metals; and should its basis ever be separated in the pure form, it will probably be one of the most powerful agents in chemistry.

I have lately made several new attempts to procure uncombined dry muriatic acid; but they have been all unsuccessful.

I heated intensely, in an iron tube, silex in a very minute state of division, and muriate of soda that had been fused; but there was not the smallest quantity of gas evolved. In this case, the silex had been ignited to whiteness before it was used; but when silex in its common state was employed, or when aqueous vapour was passed over a mixture of dry silex and dry salt in a porcelain tube, muriatic acid gas was developed with great rapidity.

I have stated page 79, that a sublimate is formed by the combustion of the olive-coloured oxide of boracium in oxymuriatic acid. On the idea that this might be boracic acid, and that dry muriatic acid might be separated in the process, I examined the circumstances of the experiment; but I found the sublimate to be a compound of boracic and muriatic acid, similar to the compound of muriatic and phosphoric acid.

I heated freshly sublimed muriate of ammonia with potassium; when the quantities were equal, as much hydrogene gas was developed as is generated by the action of water on potassium; much ammonia was evolved, and muriate of potash formed; when the potassium was to the muriate as 4 to 1, less hydrogen appeared, and a triple compound of muriatic acid, ammonia, and potassium, or its protoxide was formed, which was of a dark gray colour, and gave ammonia and muriate of potash by the action of water. There was not the slightest indications of the decomposition of the acid in the experiment. The process, in which this decomposition may be most reasonably conceived to take place, is in the combustion of potassium in the phosphuretted muriatic acid, deprived by simple distillation with potassium of as much phosphorus as possible. I am preparing an apparatus for performing this experiment, in a manner which, I hope, will lead to distinct conclusions.

The Bakerian Lecture for 1809.　.　.　.

Read November 16, 1809.

*V. Some Considerations of Theory illustrated by new Facts.**

*　　*　　*　　*　　*

Muriatic acid gas, as I have shewn, and as is further proved by the researches of MM. GAY LUSSAC and THENARD, is a compound of a body unknown in a separate state, and water. The water, I believe, cannot be decompounded, unless a new combination is formed ; thus it is not changed by charcoal ignited in the gas by

* [From " Philosophical Transactions " for 1810, vol. 100, p. 67.]

VOLTAIC electricity; but it is decompounded by all the metals; and in these cases hydrogene is elicited, in a manner similar to that in which one metal is precipitated by another; the oxygene being found in the new compound. This at first view, might be supposed in favour of the idea, that hydrogene is a simple substance; but the same reasoning may be applied to a protoxide as to a metal; and in the case of the nitromuriatic acid, when the nitrous acid is decomposed to assist in the formation of a metallic muriate, the body disengaged (nitrous gas,) is known to be in a high state of oxygenation.

* * * * *

Researches on the oxymuriatic Acid, its Nature and Combinations; and on the Elements of the muriatic Acid. With some Experiments on Sulphur and Phosphorus, made in the Laboratory of the Royal Institution. * †

Read July 12, 1810.

THE illustrious discoverer of the oxymuriatic acid considered it as muriatic acid freed from hydrogene,‡ and the common muriatic acid as a compound of hydrogene and oxymuriatic acid; and on this

---

* Communicated to the Royal Society at the request of the Managers of the Royal Institution.

† [From "Philosophical Transactions" for 1810, vol. 100, pp. 231-257. Pp. 251–257 are not reprinted.]

‡ Mem. Acad. Stockholm for 1774, p. 94.

theory he denominated oxymuriatic acid dephlogisticated muriatic acid.

M. BERTHOLLET,[*] a few years after the discovery of SCHEELE, made a number of important and curious experiments on this body; from which he concluded, that it was composed of muriatic acid gas and oxygene; and this idea for nearly twenty years has been almost universally adopted.

Dr. HENRY, in an elaborate series of experiments, made with the view of decomposing muriatic acid gas, ascertained that hydrogene was produced from it, by electricity; and he attributed the phænomenon to water contained in the gas.[†]

In the Bakerian lecture for 1808, I have given an account of the action of potassium upon muriatic acid gas, by which more than one-third of its volume of hydrogene is produced; and I have stated, that muriatic acid can in no instance be procured from oxymuriatic acid, or from dry muriates, unless water or its elements be present.

In the second volume of the Memoires d'Arcueil, M. M. GAY LUSSAC and THENARD have detailed an extensive series of facts upon muriatic acid, and oxymuriatic acid. Some of their experiments are similar to those I have detailed in the paper just referred to; others are peculiarly their own, and of a very curious kind: their general conclusion is, that muriatic acid gas contains about one quarter of its weight of water; and that oxymuriatic acid is not decomposable by any substances but hydrogene, or such as can form triple combinations with it.

One of the most singular facts that I have observed on this subject, and which I have before referred to, is,

---

[*] Journal de Physique, 1785, p. 325.
[†] Phil. Trans. for 1800, p. 191.

that charcoal, even when ignited to whiteness in oxy-muriatic or muriatic acid gases, by the VOLTAIC battery, effects no change in them; if it has been previously freed from hydrogene and moisture by intense ignition in vacuo.

This experiment, which I have several times repeated led me to doubt of the existence of oxygene in that substance, which has been supposed to contain it above all others in a loose and active state; and to make a more rigorous investigation than had been hitherto attempted for its detection.

If oxymuriatic acid gas be introduced into a vessel exhausted of air, containing tin; and the tin be gently heated, and the gas in sufficient quantity, the tin and the gas disappear, and a limpid fluid, precisely the same as Libavius's liquor is formed;—it occured to me, that if this substance is a combination of muriatic acid and oxide of tin, oxide of tin ought to be separated from it by means of ammonia. I admitted ammoniacal gas over mercury to a small quantity of the liquor of Libavius; it was absorbed with great heat, and no gas was generated; a solid result was obtained, which was of a dull white colour; some of it was heated, to ascertain if it contained oxide of tin; but the whole volatilized, producing dense pungent fumes.

Another experiment of the same kind, made with great care, and in which the ammonia was used in great excess, proved that the liquor of Libavius cannot be decompounded by ammonia; but that it forms a new combination with this substance.

I have described, on a former occasion, the nature of the operation of phosphorus on oxymuriatic acid, and I have stated that two compounds, one fluid, and the other solid, are formed in the process of combustion, of which the first, on the generally received theory of the nature of oxymuriatic acid, must be considered as a compound of muriatic acid and phosphorous acid, and the other ot muriatic acid and phosphoric acid. It occured to me, that if the acids of phosphorus really existed in these combinations, it would not be difficult to obtain them, and thus to gain proofs of the existence of oxygene in oxymuriatic acid.

I made a considerable quantity of the solid compound of oxymuriatic acid and phosphorus by combustion, and saturated it with ammonia, by heating it in a proper receiver filled with ammoniacal gas, on which it acted with great energy, producing much heat; and they formed a white opaque powder. Supposing that this substance was composed of the dry muriates and phosphates of ammonia; as muriate of ammonia is very volatile, and as ammonia is driven off from phosphoric acid, by a heat below redness, I conceived that by igniting the product obtained, I should procure phosphoric acid; I therefore introduced some of the powder into a tube of green glass, and heated it to redness, out of the contact of air by a spirit lamp; but found, to my great surprise, that it was not at all volatile nor decomposable at this degree of heat, and that it gave off no gaseous matter.

The circumstance that a substance composed principally of oxymuriatic acid, and ammonia, should resist decomposition or change at so high a temperature, induced me to pay particular attention to the properties of this new body.

It had no taste nor smell; it did not seem to be soluble, nor did it undergo any perceptible change when digested in boiling water: it did not appear to be acted upon by sulphuric, muriatic, or nitric acids, nor by a strong lixivium of potash. The only processes by which it seemed susceptible of decomposition were by combustion, or the action of ignited hydrat of potash. When brought into the flame of a spirit lamp and made red-hot, it gave feeble indications of inflammation, and tinged the flame of a yellow colour, and left a fixed acid, having the properties of phosphoric acid. When acted on by red-hot hydrat of potash, it emitted a smell of ammonia, burnt where it was in contact with air, and appeared to dissolve in the alkali. The potash which had been so acted upon gave muriatic acid, by the addition of sulphuric acid.

I heated some of the powder to whiteness, in a tube of platina; but it did not appear to alter; and after ignition gave ammonia by the action of fused hydrat of potash.

I caused ammonia, made as dry as possible, to act on the phosphuretted liquor of M. M. Gay Lussac and Thenard; and on the sulphuretted muriatic liquor of Dr. Thomson; but no decomposition took place; nor was any muriate of ammonia formed when proper precautions were taken to exclude moisture. The results were new combinations; that from the phosphoretted liquor, was a white solid, from which a part of the phosphorus was separated by heat; but which seemed no further decomposable, even by ignition. That from the sulphuretted liquor was likewise solid, and had various shades of colour, from a bright purple to a golden yellow, according as it was more or less saturated with ammonia; but as these compounds did not present the same uniform and interesting properties, as that from the phosphoric sublimate, I did not examine them minutely: I contented myself by ascertaining that no substance known to contain oxygene could be procured from oxymuriatic acid, in this mode of operation.

It has been said, and taken for granted by many chemists, that when oxymuriatic acid and ammonia act upon each other, water is formed; I have several times made the experiment, and I am convinced that this is not the case. When about 15 or 16 parts of oxymuriatic acid gas are mixed with from 40 to 45 parts of ammoniacal gas, there is a condensation of nearly the whole of the acid and alkaline gasses, and from 5 to 6 parts of nitrogene are produced; and the result is dry muriate of ammonia.

Mr. CRUIKSHANK has shown that oxymuriatic acid and hydrogene, when mixed in proportions nearly equal, produce a matter almost entirely condensible by water; and M. M. GAY LUSSAC and THENARD, have stated that this matter is common muriatic acid gas, and that no water is deposited in the operation. I have made a number of experiments on the action of oxymuriatic acid gas and hydrogene. When these bodies were mixed in equal volumes over water, and introduced into an exhausted vessel and fired by the electric spark, there was always a deposition of a slight vapour, and a condensation of from $\frac{1}{10}$ to $\frac{1}{20}$ of the volume; but the gas remaining was muriatic acid gas. I have attempted to make the experiment in a manner still more refined, by drying the oxymuriatic acid and the hydrogene by introducing them into vessels containing muriate of lime, and by suffering them to combine at common temperatures; but I have never been able to avoid a slight condensation; though in proportion as the gasses were free from oxygene or water, this condensation diminished.

I mixed together sulphuretted hydrogene in a high degree of purity and oxymuriatic acid gas both dried, in equal volumes: in this instance the condensation was not $\frac{1}{40}$; sulphur, which seemed to contain a little oxymuriatic acid, was formed on the sides of the vessel; no vapour was deposited; and the residual gas contained about $\frac{19}{20}$ of muriatic acid gas, and the remainder was inflammable.

M. M. GAY LUSSAC and THENARD have proved by a copious collection of instances, that in the usual cases where oxygene is procured from oxymuriatic acid, water is always present, and muriatic acid gas is formed; now, as it is shewn that oxymuriatic acid gas is converted into muriatic acid gas, by combining with hydrogene, it is scarcely possible to avoid the conclusion, that the oxygene is derived from the decomposition of water, and, consequently, that the idea of the existence of water in muriatic acid gas, is hypothetical, depending upon an assumption which has not yet been proved—the existence of oxygene in oxymuriatic acid gas.

M. M. GAY LUSSAC and THENARD indeed have stated an experiment, which they consider as proving that muriatic acid gas contains one quarter of its weight of combined water. They passed this gas over litharge, and obtained so much water; but it is obvious that in this case they formed the same compound as that produced by the action of oxymuriatic acid on lead; and in this process the muriatic acid must loose its hydrogene, and the lead its oxygene; which of course would form water; these able chemists, indeed, from the conclusion of their memoir, seem aware, that such an explanation may be given, for they say that the oxymuriatic acid *may be* considered as a simple body.

I have repeated those experiments which led me first to suspect the existence of combined water in muriatic acid, with considerable care; I find that, when mercury is made to act upon 1 in volume of muriatic acid gas, by VOLTAIC electricity, all the acid disappears, calomel is formed, and about .5 of hydrogene evolved.

With potassium, in experiments made over very dry mercury, the quantity of hydrogene is always from 9 to 11, the volume of the muriatic acid gas used being 20.

And in some experiments made very carefully by my

brother Mr. JOHN DAVY, on the decomposition of muriatic
acid gas, by heated tin and zinc, hydrogene equal to about
half its volume was disengaged, and metallic muriates, the
same as those produced by the combustion of tin and
zinc in oxymuriatic gas, resulted.

It is evident from this series of observations, that
SCHEELE'S view, (though obscured by terms derived from
a vague and unfounded general theory,) of the nature of
the oxymuriatic and muriatic acids, may be considered as
an expression of facts ; whilst the view adopted by the
French school of chemistry, and which, till it is minutely
examined, appears so beautiful and satisfactory, rests in
the present state of our knowledge, upon hypothetical
grounds.

When oxymuriatic acid is acted upon by nearly an
equal volume of hydrogene, a combination takes place
between them, and muriatic acid gas results.   When
muriatic acid gas is acted on by mercury, or any other
metal, the oxymuriatic acid is attracted from the hydrogene, by the stronger affinity of the metal ; and an
oxymuriate, exactly similar to that formed by combustion,
is produced.

The action of water upon those compounds, which have
been usually considered as muriates, or as dry muriates,
but which are properly combinations of oxymuriatic acid
with inflammable bases, may be easily explained, according to these views of the subject.   When water is added
in certain quantities to Libavius's liquor, a solid crystallized mass is obtained, from which oxide of tin and
muriate of ammonia can be procured by ammonia.   In
this case, oxygene may be conceived to be supplied to
the tin, and hydrogene to the oxymuriatic acid.

The compound formed by burning phosphorus in
oxymuriatic acid, is in a similar relation to water : if
that substance be added to it, it is resolved into two

powerful acids; oxygene, it may be supposed, is furnished to the phosphorus to form phosphoric acid, hydrogene to the oxymuriatic acid to form common muriatic
acid gas.

None of the combinations of the oxymuriatic acid
with inflammable bodies, can be decomposed by dry
acids; and this seems to be the test which distinguishes
the oxymuriatic combinations from the muriates, though
they have hitherto been confounded together. Muriate
of potash for instance, if M. BERTHOLLET's estimation
of its composition, approaches towards accuracy, when
ignited, is a compound of oxymuriatic acid with potassium; muriate of ammonia, is a compound of muriatic
acid gas and ammonia; and when acted on by potassium, it is decompounded; the oxymuriatic acid may be
conceived to combine with the potassium to form
muriate of potash, and the ammonia and hydrogene are
set free.

The vivid combustion of bodies in oxymuriatic acid
gas, at first view, appears a reason why oxygene should
be admitted in it; but heat and light are merely results
of the intense agency of combination. Sulphur and
metals, alkaline earths and acids become ignited during
their mutual agency; and such an effect might be expected in an operation 'so rapid, as that of oxymuriatic
acid upon metals and inflammable bodies.

It may be said, that a strong argument in favour of the
hypothesis, that oxymuriatic acid consists of an acid basis
united to oxygene, exists in the general analogy of the
compounds of oxymuriatic acid and metals, to the common
neutral salts; but this analogy when strictly investigated,
will be found to very indistinct, and even allowing it, it
may be applied with as much force to support an opposite
doctrine, namely, that the neutral salts are compounds of
bases with water; and the metals of bases with hydrogene;

and that in the case of the action of oxymuriatic acid and metals, the metal furnishes hydrogene to form muriatic acid, and a basis to produce the neutral combination.

That the quantity of hydrogene evolved during the decomposition of muriatic acid gas by metals, is the same that would be produced during the decomposition of water by the same bodies, appears, at first view, an evidence in favour of the existence of water in muriatic acid gas; but as there is only one known combination of hydrogene with oxymuriatic acid, one quantity must always be separated. Hydrogene is disengaged from its oxymuriatic combination, by a metal, in the same manner as one metal is disengaged by another, from similar combinations; and of all inflammable bodies that form compounds of this kind, except perhaps phosphorus and sulphur, hydrogene is that which seems to adhere to oxymuriatic acid with the least force.

I have caused strong explosions from an electrical jar, to pass through oxymuriatic gas, by means of points of platina, for several hours in succession; but it seemed not to undergo the slightest change.

I electrized the oxymuriates of phosphorus and sulphur for some hours, by the power of the VOLTAIC apparatus of 1000 double plates; no gas separated, but a minute quantity of hydrogene, which I am inclined to attribute to the presence of moisture in the apparatus employed; for I once obtained hydrogene from Libavius's liquor by a similar operation; but I have ascertained, that this was owing to the decomposition of water, adhering to the mercury; and in some late experiments made with 2000 double plates, in which the discharge was from platina wires, and in which the mercury used for confining the liquor was carefully boiled, there was no production of any permanent elastic matter.

As there are no experimental evidences of the exist-·
ence of oxygene in oxymuriatic acid gas, a natural
question arises, concerning the nature of these compounds, in which the muriatic acid has been supposed
to exist, combined with much more oxygene than
oxymuriatic acid, in the state in which it has been named
by Mr CHENEVIX, hyperoxygenized muriatic acid.

Can the oxymuriatic *acid* combine either with oxygene
or hydrogene, and form with each of them an acid compound ; of which that with hydrogene has the strongest,
and that with oxygene the weakest affinity for bases ? for
the able chemist to whom I have just referred, conceives
that hyperoxymuriates are decomposed by muriatic
acid. Or, is hyperoxymuriatic acid the basis of all this
class of bodies, the most simple form of this species of
matter ?

The phænomena of the composition and decomposition of the hyperoxymuriates, may be explained on either
of these suppositions ; but they are mere suppositions un-
supported by experiment.

I have endeavoured to obtain the neutralizing acid,
which has been imagined to be hyperoxygenised, from
hyperoxymuriate of potash, by various modes, but uniformly without success. By distilling the salt with dry
boracic acid, though a little oxymuriatic acid is generated,
yet oxygene is the chief gaseous product, and a muriate
of potash not decomposable is produced.

The distillation of the orange coloured fluid, produced
by dissolving hyperoxymuriate of potash in sulphuric acid,
affords only oxygene in great excess, and oxymuriatic
acid.

When solutions of muriates, or muriatic acid are
electrized in the VOLTAIC circuit, oxymuriatic acid is
evolved at the positive surface, and hydrogene at the
negative surface. When a solution of oxymuriatic acid in

· water is electrized, oxymuriatic acid and oxygene appear*
at the positive surface, and hydrogene at the negative
surface, facts which are certainly unfavourable to the idea
of the existence of hyperoxygenised muriatic acid,
whether it be imagined a compound of oxymuriatic
acid with oxygene, or the basis of oxymuriatic acid.

If the facts respecting the hyperoxymuriate of potash,
indeed, be closely reasoned upon, it must be regarded as
nothing more than as a triple compound of oxymuriatic
acid, potassium, and oxygene. We have no right to
assume the existence of any peculiar acid in it, or of a
considerable portion of combined water; and it is perhaps more conformable to the analogy of chemistry, to
suppose the large quantity of oxygene combined with
the potassium, which we know has an intense affinity
for oxygene, and which, from some experiments, I am
inclined to believe, is capable of combining directly with
more oxygene than exists in potash, than with the oxymuriatic acid, which, as far as is known, has no affinity
for that substance.

It is generally supposed that a mixture of oxymuriatic
acid and hyperoxymuriatic acid is disengaged when
hyperoxymuriate of potash is decomposed by common
muriatic acid; † but I am satisfied, from several trials,
that the gas procured in this way, when not mixed with

---

* The quantity of oxymuriatic acid in the aqueous solution, is so
small, that the principal products must be referred to the decomposition of water. This happens in other instances ; the water only
is decomposed in dilute solutions of nitric and sulphuric acids.

† If hyperoxymuriate of potash be decomposed by nitric or
sulphuric acid, it affords oxymuriatic acid and oxygene. If it be
acted upon by muriatic acid, it affords a large quantity of oxymuriatic acid gas only. In this last case, the phænomenon seems
merely to depend upon the decomposition of the muriatic acid gas,
by the oxygene, loosely combined in the salt.

oxygene, unites to the same quantity of hydrogene,* as common oxymuriatic acid gas from manganese; and I find, by a careful examination, that the gas disengaged during the solution of platina, in a mixture of nitric and muriatic acids, which has been regarded as hyperoxymuriatic acid, but which I stated some years ago to possess the properties of oxymuriatic acid gas,† is actually that body, owing its peculiar colour to a small quantity of nitromuriatic vapour suspended in it, and from which it is easily freed by washing.

Few substances, perhaps, have less claim to be considered as acid, than oxymuriatic acid. As yet we have no right to say that it has been decompounded; and as its tendency of combination is with pure inflammable matters, it may possibly belong to the same class of bodies as oxygene.

May it not in fact be a *peculiar* acidifying and dissolving principle, forming compounds with combustible bodies, analogous to acids containing oxygene, or oxides, in their properties and powers of combination; but differing from them, in being for the most part, decom-

---

* This likewise appears from Mr. CRUICKSHANK's experiments. See Nicholson's Journal, Vol. V. 4to. p. 206.

† The platina, I find by several experiments, made with great care, has no share in producing the evolution of this gas. It is formed during the production of aqua regia. The hydrogene of the muriatic acid attracts oxygen from the nitric acid. Oxymuriatic acid gas is set free, and nitrous gas remains in the solution, and gives it a deep red colour. *Nitrous* acid and muriatic acid produce no oxymuriatic acid gas. Platina, during its solution in perfectly formed aqua regia, gives only nitrous gas and nitrous vapour; and I find, that rather more oxymuriatic acid gas is produced, by heating together equal quantities of nitric acid of 1.45, and muriatic acid of 1.18, when they are not in contact with platina, than when exposed to that metal. The oxymuriatic acid gas, produced from muriatic acid by nitric acid, I find combines with about an equal volume of hydrogene by detonation.

posable by water? On this idea muriatic acid may be considered as having hydrogene for its basis, and oxymuriatic acid for its acidifying principle. And the phosphoric sublimate as having phosphorus for its basis, and oxymuriatic acid for its acidifying matter. And Libavius's liquor, and the compounds of arsenic with oxymuriatic acid, may be regarded as analogous bodies. The combinations of oxymuriatic acid with lead, silver, mercury, potassium, and sodium, in this view would be considered as a class of bodies related more to oxides than acids, in their powers of attraction.

It is needless to take up the time of this learned Society by dwelling upon the imperfection of the modern nomenclature of these substances. It is in many cases connected with false ideas of their nature and composition, and in a more advanced state of the enquiry, it will be necessary for the progress of science, that it should undergo material alterations.

It is extremely probable that there are many combinations of the oxymuriatic acid with inflammable bodies which have not been yet investigated. With phosphorus it seems capable of combining in at least three proportions; the phosphuretted muriatic acid of Gay Lussac and Thenard is the compound containing the maximum ot phosphorus. The chrystalline phosphoric sublimate, and the liquor formed by the combustion of phosphorus in oxymuriatic acid gas, disengage no phosphorus by the action of water; the sublimate, as I have already mentioned, affords phosphoric and muriatic acid; and the liquid, I believe only phosphorous acid and muriatic acid.

The sublimate from the boracic basis gives, I believe, only boracic and muriatic acid, and may be regarded as boracium acidified by oxymuriatic acid.

It is evident, that whenever an oxymuratic combination

is decomposed by water, the oxide or acid or alkali or oxidated body formed must be in the same proportion as the muriatic acid gas, as the oxygene and hydrogene must bear the same relation to each other ; and experiments upon these compounds will probably afford simple modes of ascertaining the proportions of the elements, in the different oxides, acids, and alkaline earths.

If, according to the ingenious idea of Mr. DALTON, hydrogene be considered as 1 in weight, in the proportion it exists in water, then oxygene will be nearly 7.5 ; and assuming that potash is composed of 1 proportion of oxygene, and 1 of potassium, then potash will be 48, and potassium* about 40.5 ; and from an experiment which I have detailed in the last Bakerian lecture, on the combustion of potassium in muriatic acid gas, oxymuriatic acid will be represented by 32.9, and muriatic acid gas, of course, by 33.9 ; and this estimation agrees with the specific gravity of oxymuriatic acid gas, and muriatic acid gas. From my experiments, 100 cubical inches of oxymuriatic acid gas weigh, the reductions being made for the mean temperature and pressure, 74.5 grains ; whereas by estimation they should weigh 74.6. Muriatic acid gas I find weighs, under like circumstances, in the quantity of 100 cubic inches, 39 grains ; by estimation it should weigh 38.4 grains.

It is easy from these data, knowing the composition of any dry muriate, to ascertain the quantity of oxide or of acid it would furnish by the action of water, and consequently the quantity of oxygene with which the inflammable matter will combine.†

In considering the dry muriates, as compounds of oxymuriatic acid and inflammable bodies ; the argument that I have used in the last Bakerian lecture, to

---

* Supposing potash to contain nearly 15.6 per cent. of oxygene.

† [Note not reprinted.]

shew that potassium does not form hydrate of potash by combustion, is considerably strengthened; for from the quantity of oxymuriatic acid the metal requires to produce a muriate, it seems to be shewn that it is the simplest known form of the alkaline matter. This I think approaches to an experimentum crucis. Potash made by alcohol, and that has been heated to redness, appears to be an hydrat of potash, whilst the potash formed by the combustion of potassium must be considered as a pure metallic oxide, which requires about 19 per cent. of water to convert into a hydrat.

Amongst all the known combustible bodies, charcoal is the only one which does not combine directly with oxymuriatic acid gas; and yet there is reason for believing that this combination may be formed by the intermedium of hydrogene. I am inclined to consider the oily substance produced by the action of oxymuriatic acid gas, and olefiant gas, as a ternary compound of these bodies; for they combine nearly in equal volumes; and I find that, by the action of potassium upon the oil so produced, muriate of potash is formed, and gaseous matter, which I have not yet been able to collect in sufficient quantity to decide upon its nature, is formed. Artificial camphor, and muriatic ether, as is probable from the ingenious experiments of M. GEHLEN and M. THENARD, must be combinations of a similar kind, one probably with more hydrogene, and the other with more carbon.

One of the greatest problems in œconomical chemistry, is the decomposition of the muriates of soda and potash. The solution of this problem will, perhaps, be facilitated by these new views. The affinity of potassium and sodium for oxymuriatic acid, is very strong; but so likewise is their attraction for oxygene, and the affinity of their oxides for water. The affinities of oxymuriatic acid gas for hydrogene, and of muriatic acid gas for water, are likewise of a powerful kind. Water, therefore, should be present in all cases, when it is intended to attempt to produce alkali. It is not difficult after these views to explain the decomposition of common salt, by aluminous or silicious substances, which, as it has been long known, act only when they contain water. In these cases the sodium may be conceived to combine with the oxygene of the water and with the earth, to form a vitreous compound ; and the oxymuriatic acid to unite with the hydrogene of the water, forming muriatic acid gas.

It is also easy, according to these new ideas, to explain the decomposition of salt by moistened litharge, the theory of which has so much perplexed the most acute chemists. It may be conceived to be an instance of compound affinity : the oxymuriatic acid is attracted by the lead, and the sodium combines with the oxygene of the litharge and with water to form hydrat of soda, which gradually attracts carbonic acid from the air.

As iron has a strong affinity for oxymuriatic acid, I attempted, to procure soda by passing steam over a mixture of iron filings, and muriate of soda intensely heated : and in this way, I succeeded in decomposing some of the salt : hydrogene came over ; a little hydrate of soda was formed ; and muriate of iron was produced.

It does not seem improbable, supposing the views that have been developed accurate, that by complex affinities, even potassium and sodium in their metallic form, may be procured from their oxymuriatic combinations : for this purpose the oxymuriatic acid should be attracted by one substance, and the alkaline metals by another ; and such bodies should be selected for the experiment, as would produce compounds differing considerably in degree of volatility.

I cannot conclude the subject of the application of these doctrines, without asking permission to direct the attention of the Society, to some of the theoretical relations of the facts noticed in the preceding pages.

That a body principally composed of oxymuriatic acid and ammonia, two substances which have been generally conceived incapable of existing together, should be so difficult of decomposition, as to be scarcely affected by any of the agents of chemistry, is a phænomenon of a perfectly new kind. Three bodies, two of which are permanent gases, and the other of which is con siderably volatile, form in this instance, a substance neither fusible nor volatile, at a white heat. It could not have been expected that ammonia would remain fixed at such a temperature ; but that it should remain fixed in combination with oxymuriatic acid, would have appeared incredible, according to all the existing analogies of chemistry. The experiments on which these conclusions are founded, are, however, uniform in their results : and it is easy to repeat them. They seem to shew, that the common chemical proposition, that complexity of composition is uniformly connected with facility of decomposition, is not well founded. The compound of oxymuriatic acid, phosphorus, and ammonia, resembles an oxide, such as silex, or that of columbium in its general chemical characters, and is as refractory when treated by common re-agents; and except by the effects of combustion, or the agency of fused potash, its nature could not be detected by any of the usual methods of analysis. Is it not likely, reasoning from these circumstances, that many of the substances, now supposed to be elementary, may be reduced into simpler forms of matter? And that an intense attraction, and an equilibrium of attraction, may give to a compound, containing several constituents, that refractory character, which is generally attributed to unity of constitution, or to the homogeneous nature of its parts?

Besides the compound of the phosphoric sublimate and ammonia, and the other analogous compounds which have been referred to, it is probable that other compounds of like nature may be formed of the oxides, alkalies, and earths, with the oxymuriatic combinations, or of the oxymuriatic compounds with each other; and should this be the case, the more refined analogies of chemical philosophy will be extended by these new, and as it would seem at first view, contradictory facts. For if, as I have said, oxymuriatic acid gas be referred to the same class of bodies as oxygene gas, then, as oxygene is not an acid, but forms acids by combining with certain inflammable bodies, so oxymuriatic acid, by uniting to similar substances, may be conceived to form either acids, which is the case when it combines with hydrogene, or compounds like acids or oxides, capable of forming neutral combinations, as in the instances of the oxymuriates of phosphorus and tin.

Like oxygene, oxymuriatic acid is attracted by the positive surface in VOLTAIC combinations; and on the hypothesis of the connection of chemical attraction with electrical powers, all its energies of combination correspond with those of a body supposed to be negative in a high degree.

And in most of its compounds, except those containing the alkaline metals, which may be conceived in the highest degree positive, and the metals with which it forms insoluble compounds, it seems still to retain its negative character.

*   *   *   *   *

The Bakerian Lecture. On some of the Combinations of Oxymuriatic Gas and Oxygene, and on the chemical Relations of these Principles, to inflammable Bodies.*

Read November 15, 1810.

---

1. Introduction.

IN the last communication which I had the honour of presenting to the Royal Society, I stated a number of facts, which inclined me to believe, that the body improperly called in the modern nomenclature of chemistry, *oxymuriatic acid gas*, has not as yet been decompounded; but that it is a peculiar substance, elementary as far as our knowledge extends, and analogous in many of its properties to oxygene gas.

My objects in the present Lecture, are to detail a number of experiments which I have made for the purpose of illustrating more fully the nature, properties, and combinations of this substance, and its attractions for inflammable bodies, as compared with those of oxygene; and likewise to present some general views and conclusions concerning the chemical powers of different species of matter, and the proportions in which they enter into union.

I have been almost constantly employed, since the last session of the Society, upon these researches, yet this time has not been sufficient to enable me to approach

---

* [From "Philosophical Transactions" for 1811, vol. 101, pp. 1-3, 12-35.]

to any thing complete in the investigation. But on subjects, important both in their connexion with the higher departments of chemical philosophy, and with the œconomical applications of chemistry, I trust that even these imperfect labours will not be wholly unacceptable.

2. *On the Combinations of Oxymuriatic Gas and Oxygene with the Metals from·the fixed Alkalies.*

The intensity of the attraction of potassium for oxymuriatic gas, is shewn by its spontaneous inflammation in that substance, and by the vividness of the combustion. I satisfied myself, by various minute experiments, that no water is separated in this operation, and that the proportions of the compound are such that one grain of potassium absorbs about 1.1 cubical inch of oxymuriatic gas at the mean temperature and pressure, and that they form a neutral compound, which undergoes no change by fusion. I used, in the experiments from which these conclusions are drawn, a tray of platina for receiving the potassium; the metal was heated in an exhausted vessel, to decompose any water absorbed by the crust of potash, which forms upon the potassium during its exposure to the atmosphere, and the gas was freed from vapour by muriate of lime. Large masses of potassium cannot be made to inflame, without heat in oxymuriatic gas. In all experiments in which I fused the potassium upon glass, the retorts broke in pieces in consequence of the violence of the combustion, and even in two instances when I used the tray of platina. If oxymuriatic gas be used, not freed from vapour, or if the potassium has been previously exposed to the air, a little moisture always separates during the process of combustion. When pure potassium, and pure oxymuriatic gas are used, the result, as I have stated, is a mere binary compound, the same as muriate of potash, that has undergone ignition.

The combustion of potassium and sodium in oxygene gas, is much less vivid than in oxymuriatic gas. From this phenomenon, and from some others, I was inclined to believe that the attraction of these metals for oxygene is feebler, than their attraction for oxymuriatic gas. I made several experiments, which proved that this is the fact ; but before I enter upon a detail of them, it will be necessary to discuss more fully than I have yet attempted, the nature of the combinations of potassium and sodium with oxygene, and of potash and soda with water.

*   *   *   *   *

I shall now resume the detail of the experiments that I have made, on the relative attractions of oxymuriatic gas and oxygene, for the metals of the fixed alkalies. I burnt a grain of potassium in oxygene gas, in a retort of green glass, furnished with a stop-cock, and heated the oxide formed, to redness, to convert it into potash : half a cubical inch of oxygene was absorbed. The retort was exhausted, and very pure oxymuriatic gas admitted. The colour of the potash instantly became white, and by a gentle heat, the whole was converted into muriate of potash : a cubical inch and $\frac{1}{8}$ of oxymuriatic gas were absorbed, and exactly half a cubical inch of oxygene generated. The barometer during this operation was at 30.3, the thermometer at 62 FAHRENHEIT. I made several experiments of the same kind, but this is the only one on which I can place entire dependence. When I attempted to use larger quantities of potassium, the retort usually broke during the cooling of the glass, and it was not possible to gain any accurate results in employing metallic trays. The potassium was spread into a thin plate, and of course was much oxidated before its admission into the retort, which rendered the absorption of oxygene a little less than it ought to have been. In the process it was heated in vacuo before the combustion, to decompose the water in the crust of potash ; for in cases when this precaution was not taken, I found that hydrat of potash sublimed, and lined the upper part of the retort, and from this the oxymuriatic gas separated water as well as oxygene.

The phenomenon of the separation of water from hydrat of potash by oxymuriatic gas, was happily exemplified in an experiment in which I introduced oxymuriatic gas to the peroxide of potassium, formed in a large retort, and in which the potassium had been covered with a considerable crust of hydrat of potash. The upper part of the retort and its neck contained a white sublimate of hydrat, which had risen in combustion, and which was perfectly opaque. As soon as the gas was admitted, it instantly became transparent from the evolution of water ; and on heating the glass in contact with the sublimate, its opacity was restored, and water driven off.

In various cases in which I heated dry potash, or mixtures of potash and the peroxide, in oxymuriatic gas, there was no separation of moisture, except when the gas contained aqueous vapour ; and the oxygene evolved in the process, when the heat was strongly raised, exactly corresponded to that absorbed by the potassium.

When muriatic acid gas was introduced to potash formed from the combustion of potassium, water was instantly formed, and oxymuriate of potassium.* I have made no accurate experiment on the proportions of muriatic acid gas decomposed by potash, but I made a very minute investigation, of the nature of the mutual decomposition of this substance, and hydrat of potash.

---

* i.e. Muriate of potash.

Ten grains of hydrat of potash were heated to redness in a tray of platina, which was carefully weighed; it was introduced into a retort which was exhausted of air, and the retort was filled with muriatic acid gas. The hydrat of potash was heated by a spirit lamp; water instantly separated in great abundance, and muriate of potash formed. A strong heat was applied till the process was completed, when the tray was taken out and weighed; it had gained $2\frac{13}{18}$ grains. A minute quantity of liquid muriatic acid was added to the muriate to ensure a complete neutralization, and the tray heated to redness: there was no additional increase of weight.

In the few experiments which I have made on the action of sodium and soda on oxymuriatic gas, the phenomena appeared precisely analogous; but sodium, as might have been expected, absorbed nearly twice as much oxymuriatic gas as potassium.

When common salt that has been ignited, is heated with potassium, there is an immediate decomposition, and by giving the mixture a red heat, pure sodium is obtained; and this process affords an easy mode, and the one I have always lately adopted for procuring that metal. No hydrogene is disengaged in this operation, and two parts of potassium I find produces rather more than one of sodium.

From the series of proportions that I have communicated in my last paper, it is evident that 1 grain of potassium ought to absorb 1.08 cubical inches of oxymuriatic acid; and that the potash formed from one grain of potassium ought to decompose about 2.16 cubical inches of muriatic acid gas; and these estimations agree very nearly with the result of experiments.

The estimation of the composition of soda, as deduced from the experiments in the last Bakerian lecture, is 25.4 of oxygene to 74.6 of metal, and this would give the number representing the proportion in which sodium combines with bodies 22.; * from which it is evident, that a grain of sodium ought to absorb nearly 2 cubical inches of oxymuriatic gas, and that the same quantity converted into soda, would decompose nearly four cubical inches of muriatic gas. Muriate of soda ought on this idea to contain one proportion of sodium, 22., and one of oxymuriatic gas 32.9; and this estimation is very near that which may be gained from Dr. MARCET's analysis of this substance. Hydrat of potash ought to consist of 1 proportion of potash, represented by 48., and one of water, represented by 8.5. This gives its composition as 15.1 of water, and 84.9 of potash. Hydrat of soda ought, according to theory, to contain 1 proportion of soda 29.5, and 1 of water 8.5, which will give in 100 parts 22.4 of water; and the experiments that I have detailed, conform as well as can be expected with these conclusions.

The proportions of potash and soda indicated, in different neutral combinations, by these estimations, will be found to agree very nearly with those derived from the most accurate analysis, particularly those of M. BERTHOLLET; or the differences are such as admit of an easy explanation.

I stated in my last communication, the probability that the oxygene in the hyperoxymuriate of potash was in triple combination with the metal and oxymuriatic gas; the new facts respecting the peroxide confirm this idea. Potassium, perfectly saturated with oxygene, would probably contain six proportions; for, according to Mr. CHENEVIX's analysis, which is confirmed by one made in the Laboratory of the Royal Institution, by Mr. E. DAVY, hyperoxymuriate of potash must consist of 40.5 potassium, 32.9 oxymuriatic gas, and 45 of oxygene.

I have mentioned, that by strongly heating the per-

* [Note not reprinted.]

oxide of potassium in oxymuriatic acid, all the oxygene is expelled, and a mere combination of oxymuriatic gas and potassium formed. I thought it possible, that at a low temperature, a combination might be effected, and I have reason to believe that this is the case. I made a peroxide of potassium, by heating potassium with about twice the quantity of nitre, and admitted oxymuriatic gas which was absorbed: some oxygene was expelled on the fusion of the peroxide, but a salt remained, which gave oxymuriatic gas, as well as muriatic acid, by the action of sulphuric acid.

It seems evident, that in the formation of the hyperoxymuriate of potash, one quantity of potash is decomposed by the attraction of oxymuriatic gas to form muriate of potash; but the oxygene, instead of being set free in the nascent state, enters into combination with another portion of potash, to form a peroxide, and with oxymuriatic gas.

The proportions required for these changes may be easily deduced from the data which have been stated in the preceding pages. 5 proportions of potash, equal to 240 grains, must be decomposed to form with an equal number of proportions of oxymuriatic gas equal to 164.5 grains, 5 proportions of muriate of potash equal to 367 grains; and 5 of oxygene equal to 37.5 grains, combined with one of potash, equal to 48, must unite in triple union with one of oxymuriatic gas equal to 32.9, to form one proportion, equal to 118.4 grains of hyperoxymuriate of potash.

3. On the Combinations of the Metals of the Earths, with Oxygene and Oxymuriatic Gas.

The muriates of baryta, lime, and strontia, after being a long time in a white heat, are not decomposable by any simple attractions : thus, they are not altered by dry boracic acid, though, when ˙ water is added to them, they readily afford muriatic acid and their peculiar earths.

From this circumstance, I was induced to believe that these three compounds consist merely of the peculiar metallic bases, which I have named barium, strontium, and calcium, and oxymuriatic gas ; and such experiments as I have been able to make, confirm the conclusion.

When baryta, strontia, or lime, is heated in oxymuriatic gas to redness, a body precisely the same as a dry muriate is formed, and oxygene is expelled from the earth. I have never been able to effect so complete a decomposition of these earths by oxymuriatic gas, as to ascertain the quantity of oxygene produced from a given quantity of earth. But in three experiments made with great care I found that one of oxygene was evolved for every two in volume of oxymuriatic gas absorbed.

I have not yet tried the experiment of acting upon oxymuriatic gas by the bases of the alkaline earths ; but I have not the least doubt that these bodies would combine directly with that substance, and form dry muriates.

In the last experiments that I made on the metallization of the earths by amalgamation, I paid particular attention to the state of the products formed, by exposing the residuum of amalgams to the air. I found that baryta formed in this way was not fusible at an intense white heat, and that strontia and lime so formed gave off no water when ignited. Baryta made from chrystals of the earth, as M. BERTHOLLET has shewn, is a fusible hydrat, and I found that this earth gave moisture when decomposed by oxymuriatic gas ; and the lime, in hydrat of lime, was much more rapidly decomposed by oxymuriatic gas than quicklime, its oxygene being rapidly expelled with the water.

Some dry quicklime was heated in a retort, filled with muriatic acid gas ; water was instantly formed in great abundance, and it can hardly be doubted, that this arose from the hydrogene of the acid combining with the oxygene of the lime.

As potassium so readily decomposes common salt, I thought it might possibly decompose muriate of lime, and thus afford easy means of procuring calcium. The rapidity with which muriate of lime absorbs water, and the difficulty of freeing it even by a white heat from the last portions, rendered the circumstances of the experiments unfavourable. I found, however, that by heating potassium strongly, in contact with the salt, in a retort of difficultly fusible glass, I obtained a dark coloured matter, diffused through a vitreous mass, which effervesced strongly with water. The potassium had all disappeared, and the retort had received a heat at which potassium entirely volatilizes. I had similar results with muriate of strontia, and (though less distinct, more potassium distilling off unaltered) with muriate of baryta. Either the bases of the earths were wholly or partially deprived of oxymuriatic gas in these processes, or the potassium had entered into triple combination with the muriates. I hope on a future occasion to be able to decide this point.

Combinations of muriatic acid gas with magnesia, alumine and silex, are all decomposed by heat, the acid being driven off, and the earth remaining free. I conjectured from this circumstance, that oxymuriatic gas would not expel oxygene from these earths, and the suspicion was confirmed by experiments. I heated magnesia, alumine, and silex to redness in oxymuriatic gas, but no change took place.

M. M. Gay Lussac and Thenard have shewn that baryta is capable of absorbing oxygene; and it seems likely, (as according to Mr. CHENEVIX's experiments, most of the earths are capable of becoming hyperoxymuriates) that peroxides of their bases must exist.

I endeavoured to combine lime with more oxygene, by heating it in hyperoxymuriate of potash, but without success, at least after this process it gave off no oxygene in combining with water. The salt, called oxymuriate of lime, made for the use of the bleachers, I found gave off oxygene by heat, and formed muriate of lime.

From the proportions which I have given in the last Bakerian lecture, but which were calculated from the analyses of sulphates, it follows that if the muriates of baryta, strontia, and lime, be regarded as containing one proportion of oxymuriatic gas, and one of metal, then they would consist of 71 * barium, 46 strontium, and 21 calcium, to 32.9 of oxymuriatic gas.

To determine how far these numbers are accurate, 50 grains of each of these muriates that had been heated to whiteness were decomposed by nitrate of silver, the precipitate was collected, washed, heated, and weighed.

The muriate of baryta, treated in this way, afforded 68 grains of horn-silver.

The muriate of strontia 85 grains.

The muriate of lime 125 grains.

From experiments to be detailed in the next section, it appears that horn-silver consists of 12 of silver to 3.9 of oxymuriatic gas, and consequently that barium should be represented by 65.1, strontium by 46.1, and calcium by 20.8.

---

* If Mr. JAMES THOMPSON's analysis of sulphate of barytes be made the basis of calculation, sulphuric acid being estimated as 36, then the number representing barium will be about 65.5

4. On the Combinations of the Common Metals, with Oxygene and Oxymuriatic Gas.

In the limits which it is usual to adopt in this lecture, it will not be possible for me to give more than an outline of the numerous experiments that I have made on the combinations of oxymuriatic gas with metals; I must confine myself to a general statement of the mode of operating, and the results. I used in all cases small retorts of green glass, containing from 3 to 6 cubical inches, furnished with stopcocks. The metallic substances were introduced, the retort exhausted and filled with the gas to be acted upon, heat was applied by means of a spirit lamp, and after cooling, the results were examined, and the residual gas analysed.

All the metals that I tried, except silver, lead, nickel, cobalt, and gold, when heated, burnt in the oxymuriatic gas, and the volatile metals with flame. Arsenic, antimony, tellurium and zinc with a white flame, mercury with a red flame. Tin became ignited to whiteness, and iron and copper to redness; tungsten and manganese to dull redness; platina was scarcely acted upon at the heat of fusion of the glass.

The product from arsenic was butter of arsenic; a dense, limpid, highly volatile fluid, a non-conductor of electricity, and of high specific gravity, and which when decomposed by water, gave oxide of arsenic and muriatic acid. That from antimony, was butter of antimony, an easily fusible and volatile solid of the colour of horn-silver, of great density, crystallizing on cooling in hexahedral plates, and giving, by its decomposition by water, white oxide.

The product from tellurium, in its sensible qualities

resembled that from antimony, and gave when acted on by water white oxide.

The product from mercury was corrosive sublimate. That from zinc was similar in colour to that from antimony, but was much less volatile.

The combination of oxymuriatic gas and iron, was of a bright brown; but having a lustre approaching to the metallic, and was iridescent like the Elba iron ore. It volatilized at a moderate heat, filling the vessel with beautiful minute crystals of extraordinary splendour, and collecting in brilliant plates, the form of which I could not determine. When acted on by water, it gave red muriate of iron.

Copper formed a bright red brown substance, fusible at a heat below redness, and becoming crystalline and semi-transparent on cooling, and which gave a green fluid, and a green precipitate by the action of water.*

The substance from manganese was not volatile at a dull red heat; It was of a deep brown colour, and by the action of water became of a brighter brown: a muriate of manganese, which did not redden litmus remained in solution; and an insoluble matter remained of a chocolate colour.†

---

* It is worth enquiry, whether the precipitate from oxymuriate of copper by water is not a hydrated submuriate, analogous in its composition to the crystalized muriate of Peru. This last I find affords muriatic acid and water by heat.

The *resin of copper* discovered by BOYLE, formed by heating copper with corrosive sublimate, probably contains only 1 proportion of oxymuriatic gas, whilst that above referred to must contain 2.

† When muriate of manganese is made by solution of its oxide in muriatic acid, a neutral combination is obtained, but this is decomposed by heat; muriatic gas flies off, and brown oxide of manganese remains. In this respect manganese appears as a link between the ancient metals and the newly discovered ones. Its muriate is de-

Tungsten afforded a deep orange sublimate, which, when decomposed by water, afforded muriatic acid, and the yellow oxide of tungsten.

Tin afforded LIBAVIUS's liquor, which gave a muriate by the action of water containing the oxide of tin, at the maximum of oxidation.

Silver and lead produced horn-silver and horn-lead, and bismuth, butter of bismuth. The absorption of oxymuriatic gas was in the following proportions for two grains of each of the metals : for arsenic 3.6 cubical inches, for antimony 3.1, for tellurium 2.4, for mercury 1.05,* for zinc 3.2, for iron 5.8, for tin 4, for bismuth 1.5, for copper 3.4, for lead .9, for silver, the absorption of volume was $\frac{9}{10}$, and the increase of weight of the silver was equivalent to $\frac{6}{10}$ of a grain.†

In acting upon metallic oxides by oxymuriatic gas, I found that those of lead, silver, tin, copper, antimony, bismuth, and tellurium, were decomposed in a heat below redness, but the oxides of the volatile metals,

---

composed like that of magnesia ; and its oxide is the only one amongst those long known, as far as my experiments have gone, which neutralizes the acid energy of muriatic acid gas, so as to prevent it in solution from affecting vegetable blues.

* The gas in these experiments was not freed from aqueous vapour, and as stopcocks of brass were used, a little gas might have been absorbed by the surface of this metal, so that the processes offer only approximations to the composition of the oxymuriates. The processes on lead, tellurium, iron, antimony, copper, tin, mercury, and arsenic, were carried on in three successive days, during which the height of the mercury in the barometer varied from 30.26 inches to 30.15, and the height of that in the thermometer from 63.5 to 61 FAHRENHEIT.

The experiment on silver was made at the temperature of 52 FAHRENHEIT, and under a pressure equal to that of 29.9 inches.

† This agrees nearly with another experiment made by my brother, Mr. JOHN DAVY, in which 12 grains of silver increased to 15.9 during their conversion into horn-silver.

more readily than those of the fixed ones. The oxides of cobalt and nickel were scarcely acted upon at a dull red heat. The red oxide of iron was not affeoted at a strong red heat, whilst the black oxide was rapidly decomposed at a much lower temperature; arsenical acid underwent no change at the greatest heat that could be given it in the glass retort, whilst the white oxide readily decomposed.

In cases where oxygene was given off, it was found exactly the same in quantity as that which had been absorbed by the metal. Thus 2 grains of red oxide of mercury absorbed $\frac{9}{10}$ of a cubical inch of oxymuriatic gas, and afforded 0.45 of oxygene.* Two grains of

---

* I have made two analyses of corrosive sublimate and calomel, with considerable care. I decomposed 100 grains of corrosive sublimate, by 90 grains of hydrat of potash. This afforded 79.5 grains of orange coloured oxide of mercury, 40 grains of which afforded 9.15 cubical inches of oxygene gas ; the muriate of silver formed from the 100 grains was 102.5.

100 grains of calomel, decomposed by 90 grains of potash, afforded 82 grains of olive coloured oxide of mercury, of which 40 grains gave by decomposition by heat 4.8 cubical inches of oxygene. The quantity of horn-silver formed from the 100 grains was 58.75 grains.

In the second analysis, the quantity of oxide obtained from corrosive sublimate was 78.7 ; the quantity of muriate of silver formed was 103.4 ; the oxide produced from calomel weighed 83 grains ; the horn-silver formed was $57\frac{1}{2}$ grains. I am inclined to put most confidence in the last analyses ; but the tenor of both is to shew that the quantity of oxymuriatic gas in corrosive sublimate, is exactly double that in calomel, and that the orange oxide contains twice as much oxygene as the black, the mercury being considered as the same in all. The olive colour of the oxide formed from calomel, is owing to a slight admixture of orange oxide, formed by the oxygene of the water used in precipitation ; the tint I find is almost black, when a boiling solution of potash is used; and trituration, with a little orange oxide brings the tint to olive. It has been stated, that the olive oxide thrown down from calomel by potash is

dark olive oxide from calomel decomposed by potash, absorbed about $\frac{94}{100}$ of oxymuriatic gas, and afforded $\frac{24}{100}$ of oxygene, and corrosive sublimate was produced in both cases.

In the decomposition of the white oxide of zinc, oxygene was expelled exactly equal to half the volume of the oxymuriatic acid absorbed. In the case of the decomposition of the black oxide of iron, and the white oxide of arsenic, the changes that occurred were of a very beautiful kind; no oxygene was given off in either case, but butter of arsenic, and arsenical acid formed in one instance, and the ferruginous sublimate, and red oxide of iron in the other.

Two grains of white oxide of arsenic absorbed o.8 of oxymuriatic gas.*

I doubt not that the same phenomena will be found to occur in other instances, in which the metal has comparatively a slight attraction only for oxymuriatic gas, and when it is susceptible of different degrees of oxydation, and in which the peroxide is used.

The only instance in which I tried to decompose a common metallic oxide, by muriatic acid, was in that of the fawn coloured oxide of tin; a compound of water and LIBAVIUS's liquor separated.

---

a submuriate; but I have never been able to find a vestige of muriatic acid in it when well washed. It is not easy to obtain perfect precision in analyses of the oxides of mercury; water adheres to the oxides, which cannot be entirely driven off without the expulsion of some oxygene. In all my experiments, though the oxides had been heated to a temperature above 212, a little dew collected in the neck of the retort, so that the 40 grains must have been over-rated.

* A singular instance of the tendency of the oxide of arsenic to become arsenical acid, occurs in its action on fused hydrat of potash, the water in the hydrat is rapidly decomposed, and arseniuretted hydrogene evolved, and arseniate of potash formed.

From the proportions which may be gained in considering the volumes of oxymuriatic gas absorbed by the different metals, in their relations to the quantity of oxygene which would be required to convert them into oxides, it would appear, that in the experiments to which I have referred, either one, two, or three proportions of oxymuriatic gas combine with one of metal, and consequently, from the composition of the muriates, it will be easy to obtain the numbers representing the proportions in which these metals may be conceived to enter into other compounds.*

5. *General Conclusions and Observations, illustrated by Experiments.*

All the conclusions which I ventured to draw in my last communication to the Society, will, I trust, be found to be confirmed by the whole series of these new enquiries.

Oxymuriatic gas combines with inflammable bodies, to form simple binary compounds; and in these cases, when it acts upon oxides, it either produces the expulsion of their oxygene, or causes it to enter into new combinations.

If it be said that the oxygene arises from the decomposition of the oxymuriatic gas, and not from the oxides, it may be asked, why it is always the quantity contained in the oxide; and why in some cases, as those of the

---

* From the experiments detailed in the note in the preceding pages, it would appear that the number representing the proportion in which mercury combines must be about 200. That of silver, as would appear from the results, page 52, about 100. The numbers of other metals may be learnt from the data in the same page, but from what has been stated, these data cannot be considered as very correct.

peroxides of potassium and sodium, it bears no relation to the quantity of gas?

If there existed any acid matter in oxymuriatic gas, combined with oxygene, it ought to be exhibited in the fluid compound of one proportion of phosphorus, and two of oxymuriatic gas ; for this, on such an assumption, should consist of muriatic acid (on the old hypothesis, free from water) and phosphorous acid ; but this substance has no effect on litmus paper, and does not act under common circumstances, on fixed alkaline bases, such as dry lime or magnesia. Oxymuriatic gas, like oxygene, must be combined in large quantity with peculiar inflammable matter, to form acid matter. In its union with hydrogene, it instantly reddens the driest litmus paper, though a gaseous body. Contrary to acids, it expels oxygene from protoxides, and combines with peroxides.

When potassium is burnt in oxymuriatic gas, a dry compound is obtained. If potassium combined with oxygene is employed, the whole of the oxygene is expelled, and the same compound formed. It is contrary to sound logic to say, that this exact quantity of oxygene is given off from a body not known to be compound, when we are certain of its existence in another ; and all the cases are parallel.

An argument in favour of the existence of oxygene in oxymuriatic gas, may be derived by some persons from the circumstances of its formation, by the action of muriatic acid on peroxides, or on hyperoxymuriate of potash ; but a minute investigation of the subject will, I doubt not, shew that the phænomena of this action are entirely consistent with the views I have brought forward. By heating muriatic acid gas in contact with dry peroxide of manganese, water I found was rapidly formed, and oxymuriatic gas produced, and the peroxide rendered brown. Now as muriatic acid gas is known to consist of oxymuriatic gas and hydrogene, there is no simple explanation of the result, except by saying that the hydrogene of the muriatic acid, combined with oxygene from the peroxide to produce water.

SCHEELE explained the bleaching powers of the oxymuriatic gas, by supposing that it destroyed colours by combining with phlogiston. BERTHOLLET considered it as acting by supplying oxygene. I have made an experiment, which seems to prove that the pure gas is incapable of altering vegetable colours, and that its operation in bleaching depends entirely upon its property of decomposing water, and liberating its oxygene.

I filled a glass globe containing dry powdered muriate of lime, with oxymuriatic gas. I introduced some dry paper tinged with litmus that had been just heated, into another globe containing dry muriate of lime ; after some time this globe was exhausted, and then connected with the globe containing the oxymuriatic gas, and by an appropriate set of stopcocks, the paper was exposed to the action of the gas. No change of colour took place, and after two days there was scarcely a perceptible alteration.

Some similar paper dried, introduced into gas that had not been exposed to muriate of lime, was instantly rendered white.*

Paper that had not been previously dried, brought in contact with dried gas, underwent the same change, but more slowly.

The hyperoxymuriates seem to owe their bleaching powers entirely to their loosely combined oxygene ; there

---

* The last experiments were made in the laboratory of the Dublin Society ; most of the preceding ones in the laboratory of the Royal Institution ; and I have been permitted to refer to them by the Managers of that useful public establishment.

is a strong tendency in the metal of those in common use, to form simple combinations with oxymuriatic gas, and the oxygene is easily expelled or attracted from them.

It is generally stated in chemical books, that oxymuriatic gas is capable of being condensed and crystallized at a low temperature; I have found by several experiments that this is not the case. The solution of oxymuriatic gas in water freezes more readily than pure water, but the pure gas dried by muriate of lime undergoes no change whatever, at a temperature of 40 below 0° of FAHRENHEIT. The mistake seems to have arisen from the exposure of the gas to cold in bottles containing moisture.

I attempted to decompose boracic and phosphoric acids by oxymuriatic gas, but without success; from which it seems probable that the attractions of boracium and phosphorus for oxygene are stronger than for oxymuriatic gas. And from the experiments I have already detailed, iron and arsenic are analogous in this respect, and probably some other metals.

Potassium, sodium, calcium, strontium, barium, zinc, mercury, tin, lead, and probably silver, antimony, and gold seem to have a stronger attraction for oxymuriatic gas than for oxygene.

I have as yet been able to make very few experiments on the combinations of the oxymuriatic compounds with each other, or with oxides. The liquor from arsenic, and that from tin, mix, producing an increase of temperature; and the phosphuretted, and the sulphuretted liquors unite with each other, and with the liquor of LIBAVIUS, but without any remarkable phenomena.

I heated lime gently in a green glass tube, and passed the phosphoric sublimate, the saturated oxymuriate of phosphorus through it, in vapour; there was a violent action with the production of heat and light, and a gray fused mass was formed, which afforded by the action of water, muriate and phosphate of lime.

I introduced some vapour from the heated phosphoric sublimate, into an exhausted retort containing dry paper tinged with litmus: the colour slowly changed to pale red. This fact seems in favour of the idea that the substance is an acid; but as some minute quantity of aqueous vapour might have been present in the receiver, the experiment cannot be regarded as decisive: the strength of its attraction for ammonia, is perhaps likewise in favour of this opinion. All the oxymuriates that I have tried, indeed form triple compounds with this alkali; but phosphorus is expelled by a gentle heat from the other compounds of oxymuriatic gas and phosphorus with ammonia, and the substance remaining in combination is the phosphoric sublimate.

6. *Some Reflections on the Nomenclature of the Oxymuriatic Compounds.*

To call a body which is not known to contain oxygene, and which cannot contain muriatic acid, oxymuriatic acid, is contrary to the principles of that nomenclature in which it is adopted; and an alteration of it seems necessary to assist the progress of discussion, and to diffuse just ideas on the subject. If the great discoverer of this substance had signified it by any simple name, it would have been proper to have recurred to it; but, dephlogisticated marine acid is a term which can hardly be adopted in the present advanced æra of the science.

After consulting some of the most eminent chemical

philosophers in this country, it has been judged most proper to suggest a name founded upon one of its obvious and characteristic properties—its colour, and to call it *Chlorine,* or *Chloric* gas.*

Should it hereafter be discovered to be compound, and even to contain oxygene, this name can imply no error, and cannot necessarily require a change.

Most of the salts which have been called muriates, are not known to contain any muriatic acid, or any oxygene. Thus Libavius's liquor, though converted into a muriate by water, contains only tin and oxymuriatic gas, and horn-silver seems incapable of being converted into a true muriate.

I venture to propose for the compounds of oxymuriatic gas and inflammable matter, the name of their bases, with the termination *ane.* Thus argentane may signify horn-silver; stannane, Libavius's liquor; antimonane, butter of antimony; sulphurane, Dr. Thomson's sulphuretted liquor; and so on for the rest.

In cases when the proportion is one quantity of oxymuriatic gas, and one of inflammable matter, this nomenclature will be competent to express the class to which the body belongs, and its constitution. In cases when two or more proportions of inflammable matter, combine with one of gas; or two or more of gas, with one of inflammable matter, it may be convenient to signify the proportions by affixing vowels before the name, when the inflammable matter predominates, and after the name, when the gas is in excess; and in the order of the alphabet, *a* signifying two, *e* three, *i* four, and so on.

The name muriatic acid, as applied to the compound of hydrogene and oxymuriatic gas, there seems to be no reason for altering. And the compounds of this body with oxides should be characterised in the usual manner, and as the other-neutral salts.

---

* From χλωρος.

Thus muriate of ammonia and muriate of magnesia, are perfectly correct expressions.

I shall not dwell any longer at present upon this subject.—What I have advanced, I advance merely as suggestion, and principally, for the purpose of calling the attention of philosophers to it.* As chemistry improves, many other alterations will be necessary ; and it is to be

---

* It may be conceived that a name may be found for the oxymuriatic gas in some modification of its present appellation which may harmonize with the new views, and which may yet signify its relation to the muriatic acid, such as demuriatic gas, or oxymuric gas ; but in this case it would be necessary to call the muriatic acid hydrogenated muriatic acid, or hydromuriatic acid ; and the salts which contain it hydrogenated muriates or hydro-muriates ; and on such a plan, the compounds of oxymuriatic gas must be called demuriates or oxymuriates, which I conceive would create more complexity and difficulty in unfolding just ideas on this department of chemical knowledge than the methods which I have ventured to propose. It may however be right, considering the infant state of the investigation, to suspend, for a time, the adoption of any new terms for these compounds. It is possible that oxy-muriatic gas may be compound, and that this body and oxygene may contain some common principle; but at present we have no more right to say that oxymuriatic gas contains oxygene than to say that tin contains hydrogene ; and names should express things and not opinions ; and till a body is decompounded, it should be considered as simple.

In the last number of Mr. NICHOLSON's Journal, which appeared February 1st, whilst this sheet was correcting for the press, I have seen an ingenious paper, by Mr. MURRAY, of Edinburgh, in which he has attempted to shew, that oxymuriatic gas contains oxygene. His methods are, by detonating oxymuriatic gas in excess, with a mixture of hydrogene, and gaseous oxide of carbone, when he *supposes* carbonic acid is formed ; and by mixing oxymuriatic gas in excess, with sulphuretted hydrogene, when he *supposes* sulphuric acid, or sulphureous acid is formed. In some experiments, in which my brother, Mr. JOHN DAVY, was so good as to co-operate, made over boiled mercury, we found, that 7 parts of hydrogene, 8 parts of gaseous oxide of carbone, and 20 parts of oxymuriatic gas, exploded by the electric spark, diminished to about 30 measures ;

hoped that whenever they take place, they will be made independent of all speculative views, and that new names will be derived from some simple and invariable property, and that mere arbitrary designations will be employed, to signify the class to which compounds or simple bodies belong.

.and calomel was formed on the sides of the tube. On adding dry ammonia in excess, and exposing the remainder to water, a gas remained which equalled more than 9 measures, and which was gaseous oxide of carbone, with no more impurity than might be expected from the air in the gasses, and the nitrogene expelled from the ammonia ; so that the oxygene in Mr. Murray's carbonic acid, it seems, was obtained from *water*, or from the carbonic oxide. Sulphuretted hydrogene, added over dry mercury, to oxymuriatic gas in excess, inflamed in two or three experiments ; muriatic acid gas containing the vapour of oxymuriate of sulphur, was formed, which, when neutralized by ammonia, gave muriate of ammonia, and a combination of ammonia, and oxymuriate of sulphur.

When a mixture of oxymuriatic gas in excess, and sulphuretted hydrogene, was suffered to pass into the atmosphere, the smell was that of oxymuriate of sulphur ; there was not the slightest indication of the presence of any sulphuric or sulphureous acid. If Mr. Murray had used ammonia, instead of water, for analyzing his results, I do not think he would have concluded, that oxymuriatic gas is capable of decomposition by such methods.

I shall not, at present, enter upon a detail of other experiments which I have made on this subject, in co-operation with my brother, as it is his intention to refer to them, in an answer to Mr. Murray's paper.

I shall conclude, by saying, that this ingenious chemist, has mistaken my views, in supposing them hypothetical ; I merely state what I have seen, and what I have found. There *may* be oxygene in oxymuriatic gas ; but I can find none. I repeated Mr. Murray's experiments with great interest ; and their results, when *water* is excluded, entirely confirm all my ideas on the subject, and afford no support to the hypothetical ideas, which he has laboured so zealously to defend.

On a Combination of Oxymuriatic Gas and Oxygene Gas.*

Read February 21, 1811.

I SHALL beg permission to lay before the Society the account of some experiments on a compound of oxymuriatic gas and oxygene gas, which, I trust, will be found to illustrate an interesting branch of chemical enquiry, and which offer some extraordinary and novel results.

I was led to make these experiments in consequence of the difference between the properties of oxymuriatic gas prepared in different modes; it would occupy a great length of time to state the whole progress of this investigation. It will, I conceive, be more interesting that I should immediately refer to the facts; most of which have been witnessed by Members of this Body, belonging to the Committee of Chemistry of the Royal Institution.

The oxymuriatic gas prepared from manganese, either by mixing it with a muriate and acting upon it by sulphuric acid, or by mixing it with muriatic acid, is when the oxide of manganese is pure, and, whether collected over water or mercury, uniform in its properties; its colour is a pale yellowish green; water takes up about twice its volume; and scarcely gains any colour; the metals burn in it readily; it combines with hydrogene without any deposition of moisture: it does not act on nitrous gas or muriatic acid, or carbonic oxide, or sulphureous gasses,

---

* [From "Philosophical Transactions" for 1811 vol. 101, pp. 155-162.]

when they have been carefully dried. It is the substance which I employed in all the experiments on the combinations of oxymuriatic gas described in my last two papers.

The gas produced by the action of muriatic acid on the salts which have been called hyperoxymuriates, on the contrary, differs very much in its properties, according as the manner in which it is prepared and collected is different.

When much acid is employed to a small quantity of salt, and the gas is collected over water, the water becomes tinged of a lemon colour; but the gas collected is the same as that procured from manganese.

When the gas is collected over mercury, and is procured from a weak acid, and from a great excess of salt by a low heat, its colour is a dense tint of brilliant yellow green, and it possesses properties entirely different from the gas collected over water.

It sometimes explodes during the time of its transfer from one vessel to another, producing heat and light, with an expansion of volume; and it may be always made to explode by a very gentle heat, often by that of the hand.*

It is a compound of oxymuriatic gas and oxygene, mixed with some oxymuriatic gas. This is proved by the results of its spontaneous explosion. It gives off, in

---

* My brother, Mr. J. DAVY, from whom I receive constant and able assistance in all my chemical enquiries, had several times observed explosions, in transferring the gas from hyperoxymuriate of potash, over mercury, and he was inclined to attribute the phænomenon to the combustion of a thin film of mercury, in contact with a globule of gas. I several times endeavoured to produce the effect, but without success, till an acid was employed for the preparation of the gas, so diluted as not to afford it without the assistance of heat. The change of colour and expansion of volume, when the effect took place, immediately convinced me, that it was owing to a decomposition of the gas.

this process, from $\frac{1}{8}$ to $\frac{2}{8}$ its volume of oxygene, loses its vivid colour, and becomes common oxymuriatic gas.

I attempted to obtain the explosive gas in a pure form, by applying heat to a solution of it in water; but in this case, there was a partial decomposition; and some oxygene was disengaged, and some oxymuriatic gas formed. Finding that in the cases when it was most pure, it scarcely acted upon mercury, I attempted to separate the oxymuriatic gas with which it is mixed, by agitation in a tube with this metal; corrosive sublimate formed, and an elastic fluid was obtained, which was almost entirely absorbed by $\frac{1}{4}$ of its volume of water.

This gas in its pure form is so easily decomposable that it is dangerous to operate upon considerable quantities.

In one set of experiments upon it, a jar of strong glass, containing 40 cubical inches, exploded in my hands with a loud report, producing light; the vessel was broken, and fragments of it were thrown to a considerable distance.

I analysed a portion of this gas, by causing it to explode over mercury in a curved glass tube, by the heat of a spirit lamp.

The oxymuriatic gas formed, was absorbed by water; the oxygene was found to be pure, by the test of nitrous gas.

50 parts of the detonating gas, by decomposition, expanded so as to become 60 parts. The oxygene, remaining after the absorption of the oxymuriatic gas, was about 20 parts. Several other experiments were made, with similar results. So that it may be inferred, that it consists of 2 in volume of oxymuriatic gas, and 1 in volume of oxygene; and the oxygene in the gas is condensed to half its volume. Circumstances conformable to the laws of combination of gaseous fluids, so ably illustrated

E

by M. Gay Lussac, and to the theory of definite proportions.

I have stated on a former occasion, that approximations to the numbers representing the proportions in which oxygene and oxymuriatic gas combine, are found in 7.5 and 32.9. And this compound gas contains nearly these quantities.*

The smell of the pure explosive gas somewhat resembles that of burnt sugar, mixed with the peculiar smell of oxymuriatic gas. Water appeared to take up eight or ten times its volume; but the experiment was made over mercury, which might occasion an error, though it did not seem to act on the fluid. The water became of a tint approaching to orange.

When the explosive gas was detonated with hydrogene, equal to twice its volume, there was a great absorption, to more than $\frac{1}{3}$, and solution of muriatic acid was formed; when the explosive gas was in excess, oxygene was always expelled, a fact demonstrating the stronger attraction of hydrogene for oxymuriatic gas than for oxygene.

I have said that mercury has no action upon this gas in its purest form at common temperatures. Copper and antimony, which so readily burn in oxymuriatic gas, did not act upon the explosive gas in the cold: and when

---

* In page 245 of the Phil. Trans. for 1810 [p. 35], I have mentioned that the specific gravity of oxymuriatic gas, is between 74 and 75 grains per 100 cubical inches. The gas that I weighed, was collected over water and procured from hyperoxymuriate of potash, and at that time I conceived, that this elastic fluid did not differ from the oxymuriatic gas from manganese, except in being purer. It probably contained some of the new gas; for I find that the specific gravity of pure oxymuriatic gas from manganese, and muriatic acid is to that of common air, as 244 to 100. Taking this estimation, the specific gravity of the new gas will be about 238, and the number representing the proportion in which oxymuriatic gas combines, from this estimation, will be rather higher than is stated above.

they were introduced into it, being heated, it was instantly decomposed, and its oxygene set free ; and the metals burnt in the oxymuriatic gas.

When sulphur was introduced into it, there was at first no action, but an explosion soon took place : and the peculiar smell of oxymuriate of sulphur was perceived.

Phosphorus produced a brilliant explosion, by contact with it in the cold, and there was produced phosphoric acid and solid oxymuriate of phosphorus.

Arsenic introduced into it did not inflame ; the gas was made to explode, when the metal burnt with great brilliancy in the oxymuriatic gas.

Iron wire introduced into it did not burn, till it was heated so as to produce an explosion, when it burnt with a most brilliant light in the decomposed gas.

Charcoal introduced in it ignited, produced a brilliant flash of light, and burnt with a dull red light, doubtless owing to its action upon the oxygene mixed with the oxymuriatic gas.

It produced dense red fumes when mixed with nitrous gas, and there was an absorption of volume.

When it was mixed with muriatic acid gas, there was a gradual diminution of volume. By the application of heat the absorption was rapid, oxymuriatic gas was formed, and a dew appeared on the sides of the vessel.

These experiments enable us to explain the contradictory accounts that have been given by different authors of the properties of oxymuriatic gas.

That the explosive compound has not been collected before, is owing to the circumstance of water having been used for receiving the products from hyperoxymuriate of potash, and unless the water is highly saturated with the explosive gas, nothing but oxymuriatic gas is obtained ; or to the circumstance of too dense an acid having been employed.

This substance produces the phænomena which Mr. CHENEVIX, in his able paper on oxymuriatic acid, referred to the hyperoxygenised muriatic acid ; and they prove the truth of his ideas respecting the possible existence of a compound of oxymuriatic gas, and oxygene in a separate state.

The explosions produced in attempts to procure the products of hyperoxymuriate of potash by acids are evidently owing to the decomposition of this new and extraordinary substance.

All the conclusions which I have ventured to make respecting the undecompounded nature of oxymuriatic gas, are, I conceive, entirely confirmed by these new facts.

If oxymuriatic gas contained oxygene, it is not easy to conceive, why oxygene should be afforded by this new compound to muriatic gas, which must already contain oxygene in intimate union. Though on the idea of muriatic acid being a compound of hydrogene and oxy-muriatic gas, the phænomena are such as might be expected.

If the power of bodies to burn in oxymuriatic gas depended upon the presence of oxygene, they all ought to burn with much more energy in the new compound ; but copper and antimony, and mercury, and arsenic, and iron, and sulphur have no action upon it, till it is de-composed ; and they act then according to their relative attractions on the oxygene, or on the oxymuriatic gas.

There is a simple experiment which illustrates this idea ; Let a glass vessel containing brass foil be exhausted, and the new gas admitted, no action will take place ; throw in a little nitrous gas, a rapid decomposition occurs, and the metal burns with great brilliancy.

Supposing oxygene and oxymuriatic gas to belong to the same class of bodies ; the attraction between them might be conceived very weak, as it is found to be, and they are easily separated from each other, and made repulsive by a very low degree of heat.

The most vivid effects of combustion known, are those produced by the condensation of oxygene or oxymuriatic gas ; but in this instance, a violent explosion with heat and light are produced by their separation, and expansion. a perfectly novel circumstance in chemical philosophy.

This compound destroys dry vegetable colours, but first gives them a tint of red. This and its considerable absorbability by water would incline one to adopt Mr. CHENEVIX's idea that it approaches to an acid in its nature. It is probably combined with the peroxide of potassium in the hyperoxymuriate.

That oxymuriatic gas and oxygene combine and separate from each other with such peculiar phænomena, appears strongly in favour of the idea of their being distinct, though analogous species of matter. It is certainly possible to defend the hypothesis that oxymuriatic gas consists of oxygene united to an unknown basis ; but it would be possible likewise to defend the speculation that it contains hydrogene.

Like oxygene it has not yet been decomposed ; and I sometime ago made an experiment, which like most of the others I have brought forward, is very adverse to the idea of its containing oxygene.

I passed the solid oxymuriate of phosphorus in vapour, and oxygene gas together through a green glass tube heated to redness.

A decomposition took place, and phosphoric acid was formed, and oxymuriatic gas was expelled.

Now, if oxygene existed in the oxymuriate of phosphorus, there is no reason why this change should take place. On the idea of oxymuriatic gas being undecompounded, it is easily explained. Oxygene is known to have a stronger attraction for phosphorus than oxymuriatic gas has, and consequently ought to expel it from this combination.

As the new compound in its purest form is possessed of a bright yellow green colour, it may be expedient to designate it by a name expressive of this circumstance, and its relation to oxymuriatic gas. As I have named that elastic fluid Chlorine, so I venture to propose for this substance the name Euchlorine, or Euchloric gas from ευ and χλωρος. The point of Nomenclature I am not, however, inclined to dwell upon. I shall be content to adopt any name that may be considered as most appropriate by the able chemical philosophers attached to this Society.

*** In page 48, it is stated that magnesia is not decomposed by oxymuriatic gas at a red heat. From some experiments of M. M. GAY LUSSAC, and THENARD, Bullet. de la Societ. Phil. Mai, 1810, it appears that oxygene is procured by passing oxymuriatic gas over magnesia, at a high temperature, and that a muriate indecomposable by heat is proved. They attribute the presence of this oxygene to the decomposition of the acid, but according to all analogies, it must arise from the decomposition of the earth.

*An Account of some new Experiments on the
fluoric Compounds; with some Observations
on other Objects of Chemical Inquiry.**

Read February 13, 1814.

———————◆———————

*   *   *   *   *

I have made many new experiments with the hope
of decomposing chlorine, but they have been all un-
availing; nor have I been able to gain the slightest
evidence of the existence of that oxygen which many
persons still assert to be one of its elements.

I kept sulphuret of lead for some time in fusion in
chlorine, the results were sulphurane (Dr. THOMSON's
liquor) and plumbane (muriate of lead); not an atom of
sulphate of lead was formed in the experiment, though if
any oxygen had been present, this substance might have
been expected to have been produced.

I heated plumbane (muriate of lead) in sulphurous acid
gas, and likewise in carbonic acid gas, but no change was
produced; now, if oxygen had existed either in chlorine,
or in its combination with lead, there is every reason to
believe, that the attractions of the substances concerned
in these experiments would have been such as to have
produced the insoluble and fixed salts of lead, the sulphate in the first case, and the carbonate in the
second.

I shall not enter into any discussion upon the experi-

* [From "Philosophical Transactions" for 1814, vol. 104, pp. 62-
73. Part reprinted, pp. 68-72].

ments in which water is said to be produced by the action of muriatic gas on ammonia : there is, I believe, no enlightened and candid person, who has witnessed the results of processes in which large quantities of muriate of ammonia, made by the combination of the gases in close vessels, have been distilled, without being satisfied, that there is no more moisture present, than the minute quantity which is known to exist in the compound vapours diffused through ammoniacal and muriatic acid gases, which cannot be considered either as essential to the existence of the gases, or as chemically combined with them.*

One of the first experiments that I made, with the hope of detecting oxygen in chlorine, was by acting upon it by ammonia, when I found that no water was formed, and that the results were merely muriate of ammonia and azote ;† and the driest muriate of ammonia, I find, when heated with potassium, converts it into muriate of potassa, which result would be impossible on the hypothesis of oxymuriatic gas being a compound of oxygen, for, if there was a separation of water during the formation of the muriate, the same oxygen could not be supposed to be detached in water, and yet likewise to remain so as to form part of a neutral salt.

If water had been really formed during the action of chlorine on ammonia, the result would have been a most important one : it would have proved either that chlorine or azote was a compound, and contained oxygen, or that both contained this substance ; but it would not have proved the existence of oxygen in chlorine, till it had

---

* Dr. HENRY found it very difficult to free ammonia from the aqueous vapour existing in it by hydrate of potassa, and probably the hydrated muriatic vapour which I have detected in muriatic acid gas, by a freezing mixture, is not decomposable by muriate of lime.

† Philosophical Transactions for 1810. [This Reprint, pp. 25–26.]

been shewn that the azote of the ammonia was unchanged in the operation.

Some authors continue to write and speak with scepticism on the subject, and demand stronger evidence of chlorine being undecompounded. These evidences it is impossible to give. It has resisted all attempts at decomposition. In this respect, it agrees with gold, and silver, and hydrogen, and oxygen. Persons may doubt, whether these are elementary bodies; but it is not philosophical to doubt, whether they have not been resolved into other forms of matter.

By the same mode of reasoning is that in which oxygen is conceived to exist in chlorine, any other species of matter might be supposed to form one of its constituent parts; and by multiplying words all the phenomena might be satisfactorily explained. Thus in the simple view of the formation of muriatic acid, it is said one volume of chlorine combines with one of hydrogen, and they form two volumes of muriatic acid gas. In the hypothesis of chlorine containing oxygen, it is said, the oxygen of the chlorine combines with the hydrogen to form water, and this water unites to an unknown something, or dry muriatic acid, to produce a gaseous body. If it were asserted that chlorine contained azote, oxygen, and this unknown body, then it might be said, that, in the action of hydrogen on chlorine, the azote, the oxygen and the chlorine, having all attractions for hydrogen, enter into union with it, and form a quadruple compound.

Professor BERZELIUS has lately adduced some arguments, which he conceives are in favour of chlorine being a compound of oxygen from the laws of definite proportions; but I cannot regard these arguments of my learned and ingenious friend as possessing any weight. By transferring the definite proportions of oxygen to the metals, which he has given to chlorine, the explanation becomes a simple expression of facts; and there is no general canon with respect to the multiples of the proportions in which different bodies combine. Thus azote follows peculiar laws in combining with every different body; it combines with three volumes of hydrogen, with half a volume of oxygen, with 1.2 and 1½ of the same body, and with four volumes of chlorine.

The chemists in the middle of the last century had an idea, that all inflammable bodies contained phlogiston or hydrogen. It was the glory of LAVOISIER to lay the foundations for a sound logic in chemistry, by shewing that the existence of this principle, or of other principles, should not be assumed where they could not be detected.

In all cases, in which bodies support combustion or form acids, oxygen has been supposed by the greater number of modern chemists to be present; but as there are many distinct species of inflammable bodies, so there may be many distinct species of matter which combine with them with so much energy, as to produce heat and light; and various bodies appear capable of forming acids; thus hydrogen enters into the composition of nearly as many acids as oxygen, and three bodies, namely, sulphuretted hydrogen, muriatic acid, and fluoric acid which contain hydrogen, are not known to contain oxygen. The existence of oxygen in the atmosphere, and its action in the economy of nature, and in the processes of the arts, have necessarily caused it to occupy a great portion of the attention of chemists, and, being of such importance, and in constant operation, it is not extraordinary, that a greater number of phenomena should be attributed to it, than it really produces.

In the views that I have ventured to develope, neither oxygen, chlorine, or fluorine, are asserted to be elements; it is only asserted, that, as yet, they have not been decomposed. .

As the investigation of nature proceeds, it is not improbable, that other more subtile bodies belonging to this class will be discovered, and perhaps some of the characteristic differences of those substances, which apparently give the same products by analysis, may depend upon this circumstance.

*   *   *   *   *

On the fallacy of the experiments in which water is said to have been formed by the decomposition of Chlorine.*

Read February 12, 1818.

SOME experiments have been lately communicated to the Royal Society of Edinburgh, from which it has been inferred, that water is formed during the action of muriatic acid gas on certain metals, and consequently, that chlorine is decomposed in this operation.

In repeating those experiments, I have ascertained, that the water is derived from sources not suspected by the authors, and that their conclusions are unfounded. To take up the time of the Society by long experimental details and theoretical speculations on such an occasion, will be unnecessary ; I shall therefore only transiently mention the sources of error, and demonstrate their operation by two or three examples.

When muriatic acid gas is passed through flint glass tubes heated to redness, a small quantity of water is formed by the action of the gas on the oxide of lead in

---

* [From " Philosophical Transactions" for 1818, vol. 108, pp. 169-171.]

the glass, and a smaller quantity by its action on the alkali of the glass, the process being one of double affinity, the hydrogen of the muriatic acid unites to the oxygen of the oxide, and the chlorine combines with the metals.

A copious dew was formed by passing muriatic acid gas through flint glass tubes red hot, and a less copious dew, by passing it through green glass tubes. In the first instance, the glass became opaque, and gained a pearly lustre, and a combination of chlorine and lead sublimed from the hotter into the colder part of the tube. In the second, the surface of the tube became slightly opaque, but no sublimate was formed.

When fine clean iron wire was introduced into such tubes, and made red hot, and muriatic acid gas passed over it, no particular precautions being taken to free the tubes from common air, much more water appeared; but this excess of water principally owed its existence to the combination of hydrogen disengaged from the muriatic acid gas by the iron with the oxygen of the common air. I say, *principally*, because an inappreciable quantity must have been deposited from the vapour of hydrated muriatic acid in the muriatic acid gas. This was proved by filling the whole apparatus with hydrogen in another experiment, and generating the muriatic acid gas in a retort filled with hydrogen, when the water produced was no more than might have been expected from the action of the muriatic acid gas on the oxide of lead and alkali in the glass. I give the details. Above 21 grains of the first combination of chlorine and iron were formed; the quantity of moisture collected by bibulous paper, and which was a strong acid solution of the proto-muriate of iron, amounted to less than half a grain, and of this not more than two-thirds could have been water. Now, if chlorine had been decomposed in this operation, the quantity of water ought to have been at least ten times as great.

I have shown by numerous experiments, that in the action of muriatic acid gas upon metals, hydrogen, equal in bulk to half the volume of the gas, is produced; it is therefore evident, that if water had been generated by the action of muriatic acid gas on metals, it must have been the *chlorine*, or the *metal*, or both, that were decomposed. As chlorine can be freed from much of its aqueous vapour by dry muriate of lime, which is not the case with muriatic acid gas, it offers a much more unexceptionable substance for experiments of this kind. I passed 23 cubical inches of chlorine slowly through dry muriate of lime into a flint glass tube red hot, containing a green glass tube full of iron wire; the chlorine combined with this iron wire with intense heat; the bright sublimate formed was passed through more iron wire heated to redness, so as to form a considerable quantity of the first compound of chlorine with iron, which, when examined, was found exactly the same as that produced by the action of muriatic acid gas on iron. All the products were heated strongly, and the end of the glass tube kept very cool; but *not the slightest appearance of moisture was perceptible.*

In all these experiments I was assisted by Mr FARADAY of the Royal Institution.

Muriate of ammonia is not altered by being passed through porcelain or glass tubes heated to redness, but if metals be present, it offers similar results to muriatic acid gas. In one experiment, in which muriate of ammonia recently sublimed was used, instead of muriatic acid gas, the appearance of moisture was less than in the experiment on muriatic acid gas, which has been just detailed, and yet there was a considerable action on the oxide of lead in the glass, not only by the muriatic acid, but likewise by the free hydrogen of the decomposed ammonia.

Lɪsᴛ ᴏғ Pᴀᴘᴇʀs Reprinted, wholly or in part, in the present volume.

The Bakerian Lecture. An Account of some new analytical Researches on the Nature of certain Bodies, particularly the Alkalies, Phosphorus, Sulphur, Carbonaceous Matter, and the Acids hitherto undecompounded; with some general Observations on Chemical Theory. 1808, page 5

New analytical Researches on the Nature of certain Bodies, being an Appendix to the Bakerian Lecture for 1808. 1809, - - - page 18

The Bakerian Lecture for 1809. *On some new Electro-chemical Researches, on various objects, particularly the Metallic Bodies, from the Alkalies, and Earths, and on some Combinations of Hydrogene.* 1809, - - - - page 20

Researches on the oxymuriatic Acid, its Nature and Combinations; and on the Elements of the muriatic Acid. With some Experiments on Sulphur and Phosphorus, made in the Laboratory of the Royal Institution. 1810, - - page 21

The Bakerian Lecture. On some of the Combinations of Oxymuriatic Gas and Oxygene, and on the chemical Relations of these Principles to inflammable Bodies. 1810, - - - page 40

On a Combination of Oxymuriatic Gas and Oxygene Gas. 1811, - - - - - page 63

An Account of some new Experiments on the fluoric Compounds; with some observations on other Objects of Chemical Inquiry. 1814, page 71

On the fallacy of the experiments in which water is said to have been formed by the decomposition of Chlorine. 1818, - - - - page 75

ALEMBIC CLUB REPRINTS.

Crown Octavo.   Cloth.   Uniform.

May be purchased separately or in complete sets.

VOLUMES ALREADY PUBLISHED.

No. 1.—EXPERIMENTS UPON MAGNESIA ALBA, Quick-Lime and some other Alcaline Substances. By JOSEPH BLACK, M.D. 1755. 48 pp. Price 1s. 6d. net.

No. 2.—FOUNDATIONS OF THE ATOMIC THEORY: Comprising Papers and Extracts by JOHN DALTON, WILLIAM HYDE WOLLASTON, M.D., and THOMAS THOMSON, M.D. 1802-1808. 48 pp. Price 1s. 6d. net.

No. 3.—EXPERIMENTS ON AIR. Papers published in the Philosophical Transactions. By the Hon. HENRY CAVENDISH, F.R.S. 1784-1785. 52 pp. Price 1s. 6d. net.

No. 4.—FOUNDATIONS OF THE MOLECULAR THEORY: Comprising Papers and Extracts by JOHN DALTON, JOSEPH LOUIS GAY-LUSSAC, and AMEDEO AVOGADRO. 1808-1811. 52 pp. Price 1s. 6d. net.

No. 5.—EXTRACTS FROM MICROGRAPHIA. By R. HOOKE, F.R.S. 1665. 52 pp. Price 1s. 6d. net.

No. 6.—THE DECOMPOSITION OF THE ALKALIES AND ALKALINE EARTHS. Papers published in the Philosophical Transactions. By HUMPHRY DAVY, Sec. R.S. 1807-1808. 52 pp. Price 1s. 6d. net.

No. 7.—THE DISCOVERY OF OXYGEN. Part I. Experiments by JOSEPH PRIESTLEY, LL.D. 1775. 56 pp. Price 1s. 6d. net.

No. 8.—THE DISCOVERY OF OXYGEN. Part II. Experiments by CARL WILHELM SCHEELE. 1777. 46 pp. Price 1s. 6d. net.

No. 9.—THE ELEMENTARY NATURE OF CHLORINE. Papers published in the Philosophical Transactions. By HUMPHRY DAVY, Sec. R.S. 1810-1818. 80 pp. Price 2s. net.

No. 10.—RESEARCHES ON THE ARSENIATES, PHOSphates, and Modifications of Phosphoric Acid. By THOMAS GRAHAM. 1833. 46 pp. Price 1s. 6d. net.

No. 11.—ESSAYS OF JEAN REY, Doctor of Medicine, On an Enquiry into the Cause Wherefore Tin and Lead Increase in Weight on Calcination. 1630. 54 pp. Price 1s. 6d. net.

ALEMBIC CLUB REPRINTS.

Crown Octavo. Cloth. Uniform.

May be purchased separately or in complete sets.

VOLUMES ALREADY PUBLISHED.

No. 12.—THE LIQUEFACTION OF GASES. — Papers by MICHAEL FARADAY, F.R.S., 1823–1845. With an Appendix. 80 pp. Price 2s. net.

No. 13.—THE EARLY HISTORY OF CHLORINE. Papers by CARL WILHELM SCHEELE, 1774 ; C. L. BERTHOLLET, 1785 ; GUYTON DE MORVEAU, 1787 ; JOSEPH LOUIS GAY-LUSSAC and L. J. THÉNARD. 1809. 50 pp. Price 1s. 6d. net.

No. 14.—RESEARCHES ON THE MOLECULAR ASYM-metry of Natural Organic Products. Lectures by LOUIS PASTEUR. 1860. 46 pp. Price 1s. 6d. net.

No. 15.—THE ELECTROLYSIS OF ORGANIC COM-POUNDS. Papers by HERMANN KOLBE, 1845–1868. 16 pp. Price 1s. 6d.

Postage of any of the above to any part of the World, 2d. each extra.

In Preparation.

PAPERS ON ETHERIFICATION AND OTHER SUBJECTS. By ALEXANDER W. WILLIAMSON, LL.D., F.R.S., etc.

Also Published by the Alembic Club :

LECTURES ON THE HISTORY OF THE DEVELOPMENT of Chemistry since the time of Lavoisier. By Dr A. LADENBURG, Professor of Chemistry in the University of Breslau. Translated from the Second German Edition by LEONARD DOBBIN, Ph.D. Cloth, 8vo, 373. Price 6s. 6d. net ; by post, 6s. 10d.

Edinburgh :

PUBLISHED BY THE ALEMBIC CLUB.

Edinburgh Agent :

WILLIAM F. CLAY, 18 TEVIOT PLACE.

London Agents :

SIMPKIN, MARSHALL, HAMILTON, KENT & CO., LTD.

www.ingramcontent.com/pod-product-compliance
Lightning Source LLC
Chambersburg PA
CBHW020332090426
42735CB00009B/1505